Amal Benali

Simulation using artificial neural networks in geotechnical engineering

Summary

With the rapid technological and industrial development that the world has seen, particularly in construction technology with its immense oil platforms in the depths of the ocean or desert. These fabulous tourist complexes all over the world built on land of different profiles and natures, not forgetting a country's major strategic projects such as airports, hospitals, research centres and national security. This requires a suitable load-bearing structural system, capable of distributing the forces from one level to another until they reach the foot of the structure, called the foundation. The important role of this element is to transmit the service loads from the superstructure to the supporting soil layers. Such structures can only be supported by deep foundations, of which piles are the best-known form. The first step in designing a pile is to estimate the axial bearing capacity of the pile, which has been a challenge ever since the profession of geotechnical engineering was born. Several methods and approaches have been developed to overcome the uncertainty in prediction. These methods include some simplifying assumptions and/or empirical approaches. In recent years the determination of pile bearing capacity from in-situ test data as a complement to static and dynamic analysis has been used by geotechnical engineers. The Standard Penetration Test (SPT) is often the most commonly used in-situ test in countries around the world. However, the interdependence of the factors involved such as soil stratification, soil-pile interaction and the distribution of soil resistance along pile shafts implies a considerable degree of uncertainty and may preclude the performance of simple regression analyses, requiring more extensive and sophisticated approaches to ensure appropriate structural and service performance.

Alternatively, artificial neural networks (ANNs) have recently been used to predict the ultimate capacity of batteries based on in situ tests. Very recently, several researchers have used the RNAs artificial neural network approach to develop more sophisticated methods and integrated systems in conjunction with other probabilistic and evolutionary methods. Traditional back-propagation learning methods have proved to have convergence problems. This gradient search technique does not work well for obtaining a globally optimal solution in certain applications. The first part of this book is a collection and analysis of the literature on the main advances in this field between 1994 and 2018. The reader will find an overview of some of the important work carried out during this period.

TABLE OF CONTENTS:

General introduction

Piles are deep foundations that are required when the surface foundations are no longer able to ensure the proper transmission of forces from the superstructure to the soil bearing layer. These forces will be transferred to the pile by two mechanisms: one will be a distribution along the shaft of the pile in the form of a lateral force, and the other will be at the tip (tip resistance). Analysis of the load transfer mechanism has led to a considerable number of design methods for deep foundations, in particular the determination of the axial bearing capacity of piles.

Several methods and approaches have been developed to overcome the uncertainty in the prediction. These methods include some simplifying assumptions and/or empirical approaches concerning soil stratification, pile/soil interaction, and soil resistance distribution along the pile shaft...etc. The bearing capacity of piles can be determined by the following five approaches:

- Interpretation of pile loading test curves.
- Dynamic analysis methods based on wave equation analysis.
- Interpreting dynamic tests.
- Theoretical models (static analysis methods and finite elements)
- Methods based on the interpretation of in situ tests (direct and indirect methods).

Pile loading tests remain the ideal and, in some projects, necessary means of assessing the bearing capacity of piles, whereas the dynamic analysis methods applicable to driven piles are based on the analysis of the waves emanating from the hammer-pile-soil system.

Uncertainty in the measurement of the effect of the hammer interpreted through the stresses that the soil will undergo between the pile driving time and the pile loading time will cause uncertainties in the determination of the axial bearing capacity of piles.

For dynamic test methods based on the variation in acceleration and deformation during threshing. The results will be interpreted numerically. This technique appears encouraging. However, this test requires a qualified person as well as other limitations linked to the test conditions.

Theoretical methods based on static analysis or numerical models: There are considerable uncertainties concerning the horizontal coefficient of earth pressures to be considered in the analysis. In principle, the theory of bearing capacity deals with non-coherent soils. When the soil is made up of layers of non-coherent soil, a difficulty arises concerning the angle of friction to be considered. For coherent soils, the hypothesis of the superposition of effects is used to estimate the axial capacity of piles. These methods greatly underestimate the results [1]. The same applies to numerical models based on finite elements. Recently, very sophisticated finite element models have been created that adopt fairly reliable soil behaviour models [2].

In the last decade, the application of in situ testing techniques has developed considerably due to the significant improvement in testing technology (equipment and procedures). These tests have made it possible to gain a better understanding of the soil at depth and its behaviour. Bearing capacity is usually estimated on the basis of correlations with in situ tests such as the standard penetration test SPT and the static penetrometer CPT ([3];[4]). However, the interdependence of the factors involved, such as soil stratification, pile/soil interaction and the distribution of soil resistance along fut, imply a considerable level of uncertainty and become an obstacle to the implementation of simple regression (linear or non-linear). Theoretically, modelling such systems requires understanding and explicitly finding a mathematical relationship between inputs and outputs. However, these mathematical models are difficult to find. As a result, the initiation of more sophisticated approaches to ensure adequate structural and service performance is strongly advocated.

Recently, artificial neural networks (ANNs) that do not require the incorporation of assumptions or simplifications have become an effective means of solving geotechnical problems and have recently been used by researchers to predict the axial bearing capacity of piles based on in situ tests (e.g., [5]-[10]). Fig.1 shows a classification of the different types of piles generally encountered in industry.

Currently, various researchers are using artificial neural network (ANN) models to develop more efficient systems by integrating other techniques such as evolutionary methods and probabilistic techniques (e.g., [11]-[14]).

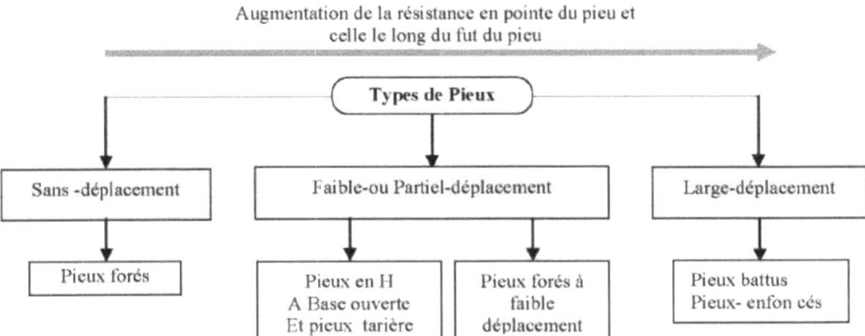

Figure 1: Classification of piles based on soil displacement during installation [15].

Aims of the study

This book presents a summary of the literature on the contributions of these techniques to geotechnical engineering and, in particular, the design of deep foundations. A historical perspective on the research carried out and a review of the application of artificial intelligence techniques and in particular artificial neural networks in geotechnical engineering are presented. As is well known, learning a neural network is an essential task in neural modelling. This requires an algorithm capable of adjusting the free properties of the network, which are the synaptic weights and the biases, and of reducing the error deduced between the calculated and experimental values according to the required performance. In this context, the backpropagation algorithm is a mathematical optimisation technique based on gradient descent for training a network. However, the various research projects carried out by the authors have revealed shortcomings in the use of this algorithm. Its convergence towards local minima is a real handicap for effective model training. Nevertheless, the researchers were able to provide developed versions of this algorithm, such as the Bayesian regularisation algorithm. The latter has demonstrated a considerable degree of success.

In recent years, there have been major contributions to the integration of new population-based heuristic algorithms, including evolutionary approaches such as genetic algorithms and teaching learning based optomization (TLBO). The second group concerns swarm intelligence methods such as particle swarm optimization (PSO) and artificial bee colony (ABC). This combination appears judicious and can lead to more accurate results.

Basic knowledge of the behaviour of axially loaded piles will then be presented. The methods for calculating the axial bearing capacity of piles based on data from the standard SPT penetration test will be discussed next. These illustrations will be followed by an assessment of the reliability and performance of methods for calculating pile bearing capacity based on the SPT test.

CHAPTER 1
Application of artificial neural networks in geotechnical engineering
1. Introduction

Artificial intelligence technology provides techniques for developing computer programmes to simulate human problem solving. These techniques include neural networks (ANNs), expert systems, fuzzy logic, etc. This discipline has techniques capable of dealing with the high complexity of geotechnical problems. The literature shows that various artificial intelligence approaches have been able to successfully solve problems such as the prediction of the load-bearing capacity of foundations, soil behaviour modelling, site characterisation, foundation settlement, slope stability, soil compaction, liquefaction, soil permeability and swelling. Artificial neural networks are one of the most widely used techniques in geotechnical engineering. The critical literature review on the use of ANNs in geotechnical engineering, and in particular in the prediction of pile bearing capacity, has shown a continuous improvement in the performance of ANN models through the more effective incorporation of optimisation methods within neural network (NN) models. A rigorous discussion of these improvements will be the subject of the next sections.

Artificial neural networks are better suited to modelling the complex behaviour encountered in most geotechnical systems, which by their very nature exhibit extreme variability. Artificial neural networks have also shown superior predictive ability when compared to traditional methods. Since the 1990s, neural networks have been successfully applied to almost all geotechnical engineering problems, and they continue to develop. In the next paragraph, a bibliographic summary of the various RNA models developed in the literature to predict the bearing capacity of axially loaded piles between 1994 and 2018 will be presented and discussed.

2. Artificial neural networks
2.1. History

The story begins in 1890, when the famous American psychologist James introduced the concept of Associative Memory [16]. Around the 1940s, cybernetics appeared, including the construction of the Turing machine. In 1943, the first Formal Neuron was produced by the neurophysiologist McCulloch and the logician Pitts, to calculate certain logical functions [17]. In 1949 Hebb produced the first Learning Rule from a psychophysiological perspective [18].

Implementing the ideas of Hebb, McCulloch and Pitts. In 1958, Rosenblatt described the first operational model of RNAs, called Perceptron, and applied it to the field of pattern recognition. [19]. In 1960, a system called Adaline was developed by automaticians Windrow and Hoff at Stanford University [20]. This system was analogous to an electronic tool made from simple components. These models gave rise to a great deal of research and probably too much hope when, in 1969, two mathematicians, Minsky and Papert, demonstrated the severe theoretical limitations of the Perceptron, in particular the impossibility of using this model to deal with non-linear problems [21]. A few years later, numerous models were implemented, but their capabilities remained limited.

After this period of stagnation, mathematics and theoretical physics were introduced into the research, which led to a real revival of RNAs, such as that of Hopfield in 1982 [22], then Grossberg [42] [23], who showed the analogy of RNAs with certain physical systems. Thanks to original contributions, this approach has led to the development of a new learning algorithm called Retropropagation [24].

In 1985, Gradient Backpropagation was introduced. It made it possible to overcome the limitations associated with the Perceptron, and to demonstrate specific properties such as parallelism, generalisation and distributed memory. This discovery made it possible to implement a non-linear input/output function on a network. A year later, Rumelhart and McClelland proposed the new architecture of RNAs and extended the use of the Retropropagation technique to various applications.

In recent years, the first practical applications of RNAs have begun to emerge, and this discipline will therefore concern an increasingly wide audience of students, researchers, engineers and industrialists, for whom the Information Society is currently demanding new applications. In France, the discipline is exemplified by the Neuro Times conference, created in 1998, on the theme of Neural Networks and their

Applications [25]. The number of participants grows every year, reflecting the interest of the scientific and industrial world (Fig. 2.1).

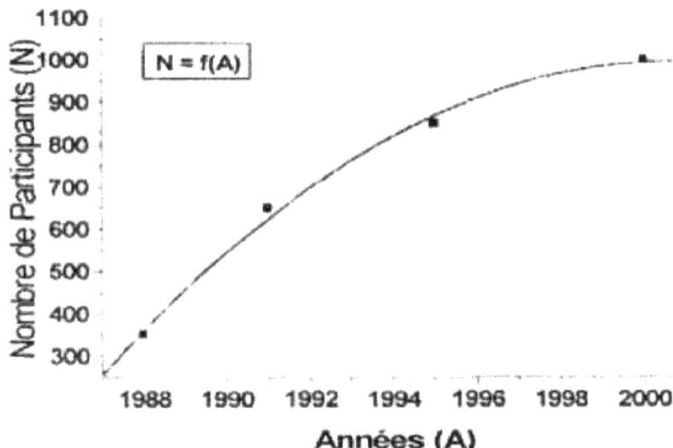

Fig.1.1 Illustration of the increase in interest in RNAs [25].

2.2. From the brain to the formal neuron

The idea of the formal neural network comes from the study of the human brain. The following is just a very schematic view of how the brain works [26]. Modelling involves finding a mathematical model that describes the behaviour of a biological neuron.

2.2.1. Biological neuron

The brain has deep roots in the past of species. It has evolved in both size and neurological complexity over several million years. It is a self-adapting system that establishes its own programmes. It is made up of around a thousand billion nerve cells known as neurons. Each of these neurons is connected to around 10,000 neighbouring neurons, giving a total number of connections of around 10^{15} . The neuron has a nucleus and extensions through which it can distribute signals via an outgoing connection (axon) or receive them via incoming connections (dendrites); signal exchanges take place at the synapse level (Fig.2.2).

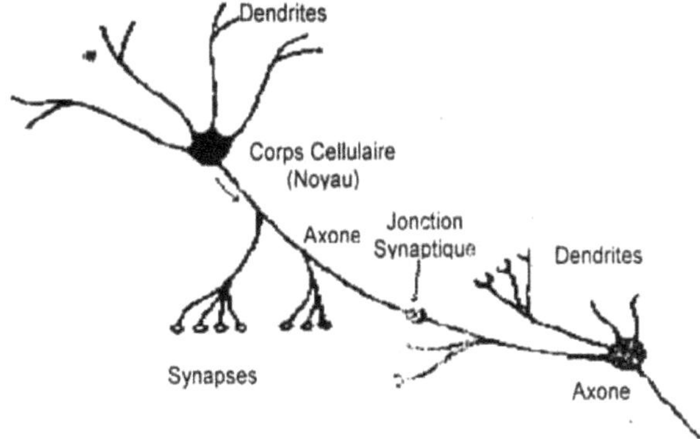

Fig.1.2. Connection between two biological neurons [27].

3.3.2. Formal neuron

The first simplified model, inspired by the biological neuron, called the formal neuron [28], is a Boolean automaton characterised by its binary state with a single output and any number of inputs. It periodically calculates the weighted sum of all the potentials available at its input. The result is then tested against a specified threshold. If the weighted sum is greater than this threshold, the neuron is activated and produces an output potential, otherwise it remains inactive and transmits nothing. It can be represented as follows (Fig.2.3)

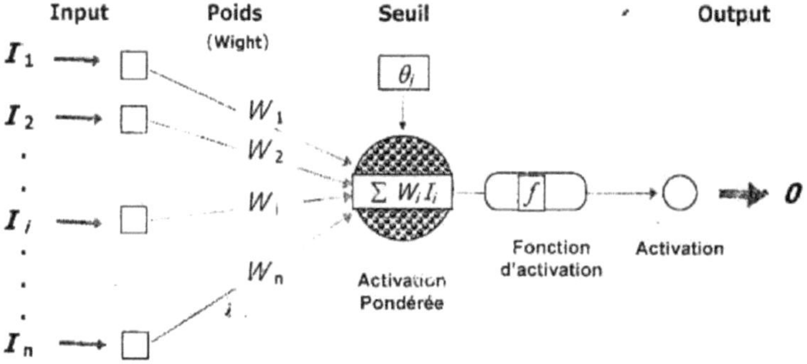

Fig.1.3. Modelling of the formal neuron [29].

The general model of a formal neuron is a very simple mathematical model derived from an analysis (also fairly simple) of biological reality. The formal neuron sums up the action potentials it receives from the other units to which it is connected, according to the expression :

$$O = f\left(\sum W_i I_i\right) \tag{2.1}$$

With : I_i, O: input and output of the neuron respectively, W_i: weight corresponding to each input; Θ: represents the threshold; and $f(x)$ threshold function verifying $\forall x$, $f(x) = 1$ si $x > \theta$, et $f(x) = 0$ otherwise. Broadly speaking, it can be defined by five elements:

-   The nature of its inputs: which may be binary or real (descriptive variables);
-   The total input function: which defines the pre-processing carried out on the inputs;
-   The neuron's activation function: which defines its internal state as a function of its total input;
-   The output function: which calculates the neuron's output as a function of its activation state; and
-   The nature of the neuron's output.

The activation (transfer) function limits the neuron's output to the minimum and maximum bounds allowed. In other words, it offers an infinite number of possible values between [0 , +1] (or [-1 , +1]). It can take different forms: binary, linear with threshold, sigmoid (also called logistic function). The choice of one of these functions is an important constituent of RNAs. Thus, identity is not always sufficient and non-linear and more advanced functions are often required. By way of illustration, here are some commonly used functions (Fig.2.4)

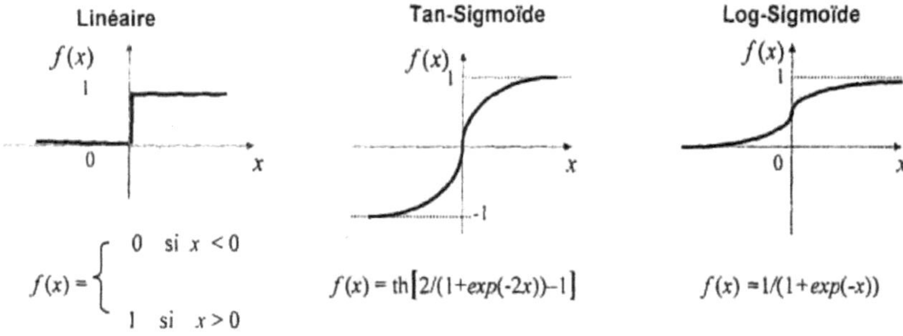

Fig.1.4 The different activation functions [29].

The artificial neuron is also characterised by another function called the base or state function. The connections of an RN are represented analytically by the function u (W, x) where x is the vector of inputs. This function describes the state of the network and generally has two forms:

-   **LBF** (Linear Basis Function), which represents a linear combination of inputs:

$$u(W,x) = \sum_{j=1}^{n} W_{ji} x_j \tag{2.2}$$

-   **NLBF** (Non Linear Basis Function) which represents a non-linear combination of inputs and can take several forms; the most commonly used is the **RBF** (Radial Basis Function) [46].

$$u(W,x) = \left(\sum_{j=1}^{n} (x_j - W_{ji})^2 \right)^{1/2} \tag{2.3}$$

2.3. Artificial neural networks

The main idea behind "modern" neural networks is as follows: We give ourselves a simple unit, a formal neuron, which is capable of performing a few elementary calculations. We then link together a large number of these units and try to determine the computing power of the neural network thus obtained. For more details, see [28-31]. Several authors are interested in neural networks, and each defines them in their own way:

According to DARPA (1988): A neural network is a system composed of several simple computing units operating in parallel, whose function is &determined by the structure of the network, the strength of the connections, and the operation performed by the elements or nodes.

According to Haykin (1994): A neural network is a massively distributed parallel processor with a natural propensity to store experiential knowledge and make it available for use. It resembles the brain in two respects:

-   Knowledge is acquired by the network through a learning process;
-   The connections between neurons (synaptic weights) are used to store knowledge.

According to Nigrin (1993): A neural network is a circuit made up of a very large number of simple computing units based on neurons. Each element operates only on local information and asynchronously, so there is no general clock for the system.

According to Zurada (1992): Artificial neural systems or neural networks are physical cellular systems that can acquire, store and use experiential knowledge.

2.4. Properties and advantages of RNAs

RNAs have some interesting properties and characteristics. They operate without a program, do not execute instructions, do not manipulate numbers, and do not contain memory to store data. The destruction of part of these circuits does not prevent the network from functioning. They are attractive for at least four reasons:

2.4.1   Parallelism: the execution of several tasks (numerical data, not symbolic data) at the same time. This is an essential feature of a high-speed neural network;

2.4.2   Generalisation: this property allows the network to find a generalised solution applicable to all examples of the problem that might be of interest, even when these examples contain errors, are incomplete or are not presented during learning;

2.4.3   Distributed memory: the memory is distributed throughout the network. The advantage of distributing the memory over several entities is its resistance to noise;

2.4.4   Learning capacity: this enables networks to take account of external constraints and data, and is characterised in some networks by their capacity for self-organisation, which ensures their stability;

2.4.5 . RNA learning mechanism

Learning is a process by which the network's free parameters (weights and biases) are adopted during the continuity of the stimulation process by the environment to which the network is reinforced. This is why most networks have a type-dependent learning rule that allows them to adapt their weights automatically. The best-known learning rules are the Hebb rule, the Windrow-Holff rule (Delta) and the generalised Delta rule (retroporopagation). The learning mechanism differs depending on the task for which the network is being used. There are three main types of learning: supervised, unsupervised and semi-supervised, as well as competitive and Boltzmann learning.

2.4.6 . Number, distribution and format of learning examples

Increasing the amount of training data provides more information about the output of the network. The distribution of training examples in the problem domain can have a significant effect on the learning and generalisation of the network. It is closely linked to the way in which the examples are produced or collected. The best-known ways of acquiring data are : synthesis of published data, -direct observation of the situation to be modelled, -consultation with a human expert, -numerical simulations. Normalisation can also be applied to the values of learning examples. This can be beneficial in cases where the order of magnitude of the examples is relatively large.

2.4.7 Perceptron model

This is a very simple neural network, with two layers, accepting only binary input and output values. Its use is limited to solving simple logical operations. The learning algorithm is the Hebb rule, in which the weights are modified by multiplying the input of a neuron by its output, thus increasing the learning rate.

2.4.8 Multilayer back-propagation model

The neurons in the input layer of the network provide the activation elements of the model (input vector), which constitute the input signals applied to the neurons in the second layer (calculation units), i.e. the first hidden layer. The output signals from the second layer are used as inputs for the third layer, and so on for the rest of the network. All the output signals from the neurons in the output layer of the network constitute the overall response of the network. A neural network is a system made up of a set of neurons interconnected with each other (Fig2.5). A certain arrangement of the connection of these neurons has produced a neural network model adapted to perform certain tasks. The multi-layer perceptron model is the most popular neural network model often used, consisting of three adjacent layers, input, hidden and output [32]. To obtain certain desired outputs, the weights, which represent the strength of connection between neurons and biases, are adjusted using a number of training inputs and the corresponding target values. The network error, i.e. the difference between the calculated and desired values, is then propagated from the output layer to the input layer to update the network weights and biases. This is the learning process.

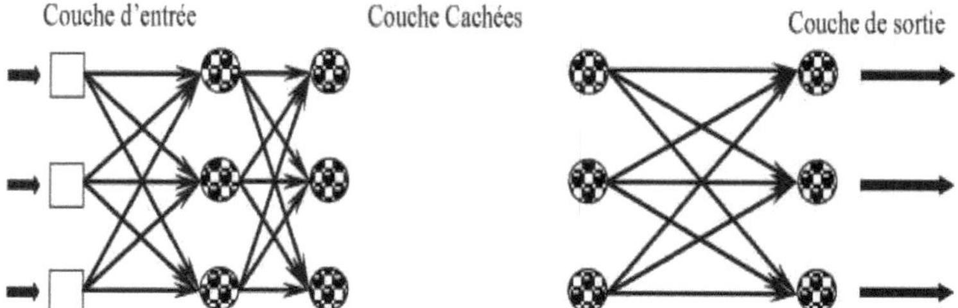

Fig 1.5. Example of a multilayer Perceptron [29].

2.4.9 Backpropagation algorithm

The backpropagation (BP) algorithm is the most widely used search technique for learning a network. The knowledge in an RNs is stored in the connection weights. The aim of training is to iteratively change the weights between neurons in a direction that minimises the error E, defined as the squared difference between the desired results and the calculated results, summed over the training dataset and the output neurons. The algorithm uses a sample-by-sample update rule to adjust the connection weights in the network. In one iteration, a training sample is presented to the network. The signal is then fed forward through the network until the output of the network is obtained. The error between the actual and desired outputs is calculated and used to adjust the connection weights. In fact, the adjustment procedure, derived from a gradient descent method, is used to reduce the error. The procedure is first applied to the connection weights in the output layer, then to the connection weights in the hidden layer next to the output layer. This calculation is continued backwards until the connection weights in the first hidden layer are reached. The iteration is completed once all the connection weights in the network have been adjusted [27].

2.5. Designing the architecture of a neural network

The structural design of an RNA involves determining the hidden layers and hidden neurons and selecting the training algorithm. Starting with the selection of the input parameters, which is a decisive step in avoiding the problem of overlearning, as there may be interdependencies and redundancies between these parameters. In this context, it is interesting to make use of statistical data analysis techniques to eliminate suspect correlations and reduce the input variables. The second part of the book will explain a statistical technique for doing this, which is PCA, and will then demonstrate its effectiveness in improving the generalisation capacity of the RNAs model developed. The other parameters, such as the number of hidden layers and the number of neurons in the hidden layers, are determined using the trial and error method. Some research work has contributed to choosing the optimal architecture and its parameters, such as the number of hidden layers, the number of neurons in each layer, the selection of the learning rate parameter, etc. ([33],[34]).

2.6. Mathematical formulation of an RNAs

When the learning process is launched, the examples are presented through the input vectors (input nodes) and their associated outputs. The data is first normalised using, for example, the function in Eq(2.4). The normalised values obtained are in the interval [-1 +1].

$$pn = 2(p - \min p)/(\max p - \min p) - 1, \quad tn = 2(t - \min t)/(\max t - \min t) - 1 \tag{2.4}$$

Where: **p** is the matrix of input vectors; **t** = is the matrix of desired output vectors; **pn** = normalised input vectors; **tn** = normalised desired output vectors; **max p** is a vector containing the maximum values of the input vectors; **min p** = is a vector containing the minimum values of the input vectors; **max t** is a vector containing the maximum value of the desired outputs and, **min t** is a vector containing the minimum value of the desired outputs.

The performance of the RNAs model is measured in terms of the MSE function (Eq 2.5) (the mean square error). The output obtained after each run (feed forward calculation) is compared with the associated desired output.

$$E = \sum_{i=1}^{Num} (T_i - t_i)^2 \tag{2.5}$$

Where, Num is the number of training data, ti is the calculated output, Ti is the desired output. Assuming that we have opted for an RNAs model with a single hidden layer, the only activation function used is the sigmoid tangent and let Y be the calculated output. The mathematical expression for the derivative model can be written as Eq(2.6 and 2.7), and the results obtained will then be denormalized (Eq2.8).

$$y = Tangsig\left(\sum y_1 . W_1 + b_1\right) \tag{2.6}$$

$$y_1 = Tangsig\left(\sum W_1 x_1 + b_0\right) \tag{2.7}$$

With :

$$t = 0.5(tn + 1)(\max t - \min t) + \min t \tag{2.8}$$

3. Review of the literature from 1994 to 2018

3.1. Axial bearing capacity of piles

The various research projects carried out to understand the behaviour of deep foundations have so far failed to find a precise solution to this complex problem. The problem is linked simultaneously to the pile and the soil into which the pile is driven. The interaction between the pile and the soil, the behaviour of the interface between these two different materials, the distribution of soil resistance along the shaft of the pile, and the interdependence between the factors involved make it impossible to implement a simple regression, and therefore require the use of more sophisticated approaches to overcome this complexity. The application of RNAs to predict the load-bearing capacity of piles began mainly in the years 1994-1995 and has undergone considerable development to improve the performance of prediction models in subsequent years.

The prediction of the axial bearing capacity of piles, particularly based on pile driving data, has been examined by several researchers. *Goh (1994a, 1995b)* [35] presented a neural network for predicting the bearing capacity due to lateral friction in clays. The multilayer perceptron model with the backpropagation algorithm was used. The model was trained with on-site experimental data. The model inputs are: pile length and diameter, principal effective stress, and undrained shear stress. The resistance due to lateral friction was chosen as the model output. The results obtained using the model were compared with those obtained by [36] and [37]. The regression analysis and the error rate were the subject of this comparison (Table 3.1). It is clear from the results that the model outperforms the other two methods.

Table 3.1 Performance of the model developed [35].

Methods	Correlation coefficient		Error level (kPa)	
	Learning	Test	Learning	Test
Model RN	0.985	0.956	1.016	1.194
Semple and Rigden (1986)	0.976	0.885	1.318	1.894
Burland (1973)	0.731	0.704	4.824	3.096

Immediately following this model, *Goh (1995a, 1996b)* [38] developed another for estimating the ultimate bearing capacity of driven piles in non-cohesive soils. The data used concern steel and reinforced concrete piles anchored in sandy soils. The inputs to the model were: the weight and drop height of the hammer, the weight - length and modulus of elasticity of the pile as well as its lateral section and the type of hammer used. The output of the model was the axial bearing capacity of the pile. The validated model was compared with other methods (Table 3.2). The values predicted by the different methods indicate that the RNs model performs better than the others.

Table 3.2 Summary of the results of the pile capacity prediction regression analysis (Goh 1995) [38].

Methods	Correlation coefficients	
	Learning	Test data
RNs model	0.96	0.97
Engineering News	0.69	0.61
Hiley	0.48	0.76
Janbu	0.82	0.89

Chan et al (1995) [39] developed a model as an alternative to the pile driving formula. The model was trained with the same parameters of the *'simplified Hiley'* formula [40], including: the respective moduli of elasticity of the pile and soil and the pile driving energy delivered to the pile. The pile bearing capacity was the sole output of the model. The desired outputs that were used in the training process were estimated using software called CAPWAP [41]. The mean square error percentage was 13.5% for the training phase and 12% for the test. The multi-layer perceptron with backpropagation algorithm has been the choice in a considerable number of geotechnical problems.

Lee and Lee (1996) [42] estimated the bearing capacity of piles using the RNs approach. Model piles were used for the simulation, using the calibration chamber and the results of full-scale pile loading tests. The inputs of the first model were: the penetration depth ratio (ratio between the penetration depth and the pile diameter), the average stress in the calibration chamber and the number of blows. The output of the model was the axial bearing capacity. The model predictions show a maximum error of no more than 20% and the mean square error is less than 15%. Note that the first model developed is based on tests on model piles in the calibration chamber, while the second model is developed with reference to full-scale pile compression tests.

For the simulation of full-scale pile loading tests, presented by the second RN model developed, five input variables were selected, the ratio of the pile penetration depth; the average number of blows along the pile and another close to the tip as well as the energy delivered by the hammer. The results obtained by the model were compared with Meyerhof's method [43]. Figure.3.1 shows that the values predicted by the RNs model are better than those given by Meyerhof's formula.

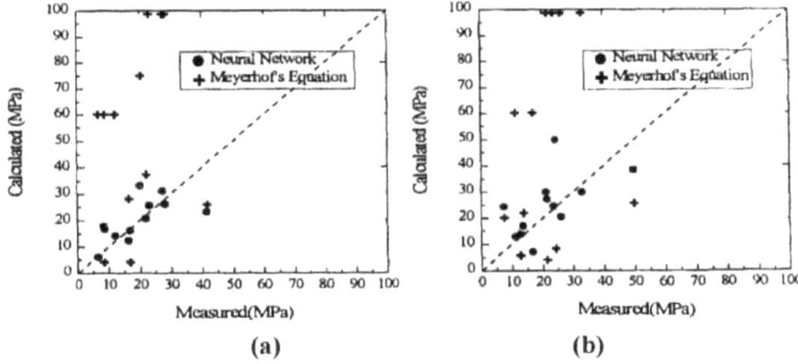

Figure 3.1 Comparison of predicted bearing capacity values with those measured using the RNs model and Meyerhof's method: **(a)** From tests in the calibration chamber, **(b)** From pile loading tests [42].

Abu-Keifa (1998) [5] developed three Generalized Regression Neural Network (GRNN) models, which is a type of probabilistic neural network (denoted; GRNNM 1, GRNNM 2 and GRNNM3) for the prediction of total axial capacity, pile tip capacity and lateral friction capacity respectively. The first model includes five input variables: the angle of internal friction of the soil around the shaft and at the base of the pile, the effective pressure due to the weight of the earth at the tip of the pile, the length of the pile and the cross-sectional area of the pile. These same variables were used for the second model. For the last model, four input variables were used: the average number of blows along the shaft, the angle of internal friction along the shaft, the length and the diameter of the pile. The results obtained were compared with four empirical techniques for determining the axial capacity of piles: Meyerhof (1976) [43], Coyle and Castello (1982) [44], API (American Petroleum Institute, 1984) [45] and Randolph (1985) [46]. The results obtained show high values for the coefficient of determination of the three models (of the order of 0.95). Figures 3.2 and 3.3 show the measured load-bearing capacities as a function of those predicted by the two models (GRNNM 1, GRNNM 2) and for the four methods chosen. The figures show that the RNs models produce values closer to the measured values than those provided by the methods.

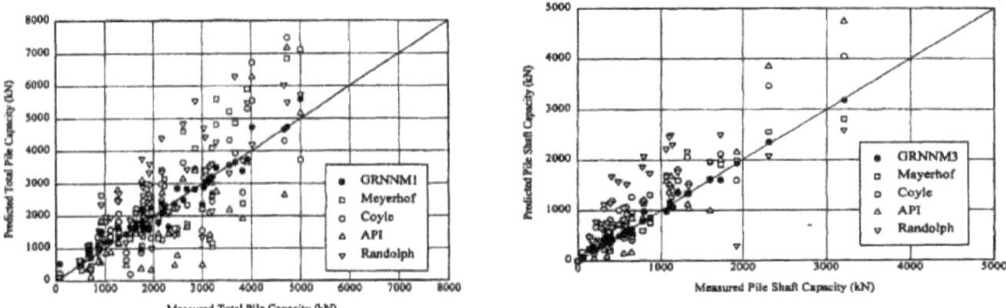

Figure 3.2 Comparison of predicted and measured total capacities [5].
Figure 3.3 Comparison of predicted lateral friction capacities with those measured [5].

Nawari et al (1999) [47] used neural networks to predict the axial load capacity of H-shaped steel, prestressed and reinforced concrete piles using the generalized regression neural network (GRNN) and backpropagation neural network (BPNN) models.

Park and Cho (2010) [48]applied an artificial neural network (ANN) to predict the strength of driven piles by dynamic loading tests. They collected 165 data for driven piles at various construction sites in Korea. The predictions of different resistances, peak resistance, lateral friction resistance and total resistance were carried out. The model has seven nodes in the input layer, eight nodes in the hidden layer and three nodes in the output layer (Fig.3.4). The combination of activation functions applied to the neurons in the

is tan-sigmoid and linear, respectively. The results indicate that the model is a reliable and simple predictive tool for appropriately taking into account various parameters essential for predicting the strength of driven piles (Fig.3.5).

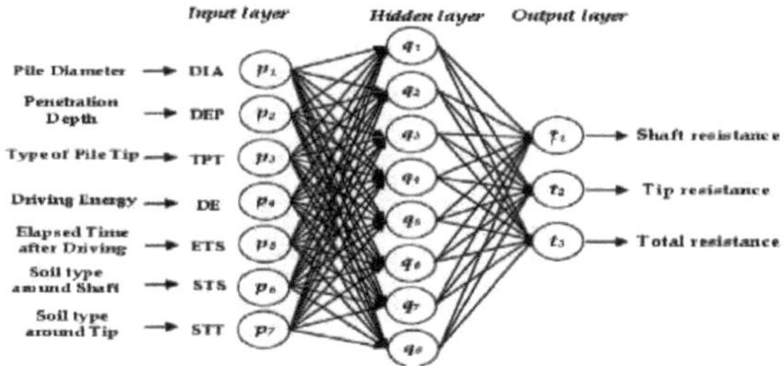

Fig 3.4 Architecture of the artificial neural network model [48]

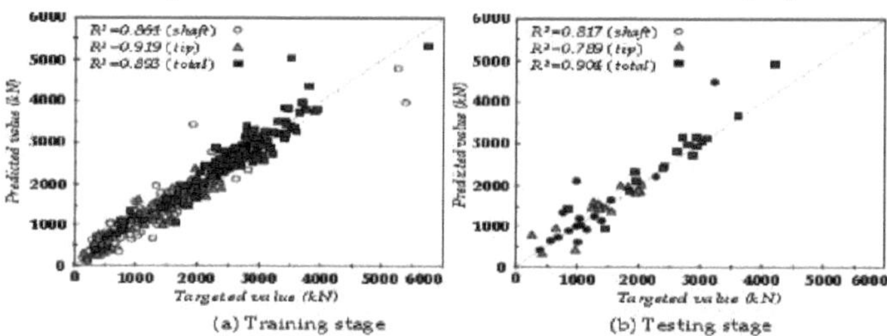

Fig 3.5. Comparison of predicted and measured pile resistance [48].

Teh et al (1997) [49] proposed an RNs model to estimate axial capacity from dynamic data based on the analysis of waves generated along the pile during driving. The desired output is derived from the CAPWAP program [50].

Kuo et al (2009) [51] have developed a RNs model that can be used to assess the bearing capacity of shallow foundations in multilayer cohesive soils. The training data used in the development of the model results from the numerical simulations carried out.

Shahin (2010) [52] developed two RNs models to predict the axial capacity of the two types of piles, driven and bored, based on the CPT test. The 80 driven piles and 94 bored piles were collected from the literature. The prediction of the models developed showed good accuracy when compared with traditional methods based on the CPT test.

Recently, *Shahin (2014)* [7], *Kordjazi et al. (2014)* [8] *and Mohammad et al. 2015* [10] have developed more reliable RN models for pile axial capacity prediction, based on in situ tests (the CPT test) and using a large database. A rigorous analysis of the data was carried out beforehand with the aim of developing more reliable RN models. The correct definition of the failure zone is an important source of the failure of the various simulation models of the behaviour of axially loaded piles. The authors therefore put emphasised this point in their studies. This simple, intelligent and fruitful manipulation has made it possible to build more competitive RN models.

Benali et al(2017) [10] developed a RNs model for the prediction of pile axial capacity based on SPT test data. The model developed is a multi-layer perceptron with Bayesian regularisation. The authors performed a statistical analysis of the training data using the principal component analysis (PCA) method. This made it possible to reduce the dimensionality of the space of input variables and eliminate any correlations and redundancies that might exist. The results were very encouraging.

3.2. Shortcomings encountered with the Backpropagation algorithm

Despite their popularity, the backpropagation (BP) algorithm has the disadvantage of converging slowly towards an optimal solution. This gradient search technique has two serious disadvantages: the model converges towards an optimal solution with inconsistent performance. The most important problem that arises during the neural network training process is the possible overfitting of the training data. In other words, over a certain training period, the network no longer improves its computing capacity. In this case, the training has become trapped in a local minimum. Among the methods suggested for trying to escape local optimums and find the global solution is the Bayesian regularisation approach. This approach minimises the over-fitting problem by taking into account the quality of the fit and the architecture of the network. It consists of modifying the objective function such as the sum of squared network errors (SSE). The objective function in Eq. (5) is extended with the addition of a term, *'Ew'* which is the sum of the squares of the network weights such that:

$$F = \beta E\alpha + \alpha Ew \tag{3.1}$$

Where: a and β are parameters that need to be optimised in this method [53].

Other clarifications

The momentum of the backpropagation algorithm is widely used to train neural networks for prediction problems (Adeli and Hung 1995) [54]. This algorithm has a very low learning level. The number of iterations to train a model is generally in the order of thousands and sometimes more [55]. However, the level of convergence depends strongly on the momuntum ratios and the learning level encountered in this algorithm. The appropriate values of these parameters depend on the type of problem ([56], [57]). Recently, other approaches have been used to overcome these shortcomings, by adopting new algorithms for learning RNAs, or combining them with other artificial intelligence techniques.

In this context, *Adeli and Hung (1994)* [58] developed a learning algorithm called *the 'conjugate adaptive gradient'* for learning the neural network, based on error minimisation. The problem of arbitrary choices of momuntum ratios and learning level encountered in the backpropagation algorithm is overcome in the new adaptive algorithm. Instead of constant learning and momuntum ratios, the authors have specified the interval in which the solution is sought during the learning process, using a mathematical approach. This algorithm is based on solid mathematical foundations, which enabled it to converge rapidly towards the solution.

Polynomial neural networks

Ardalan et al (2009) [59] have adopted an algorithm called *'Group Methods of Data Handling GMDH'*. This algorithm was developed by Ivakhnenko (1971) [60] as a multivariate analysis method for modelling complex systems without any specific knowledge of these systems. The main idea is to construct an analytical function within the neural network based on a quadratic transfer function whose coefficients are obtained using the regression technique. In their studies, the lateral capacity of the pile (due to lateral friction) from the penetrometer test (CPT) and the piezocone test (CPTu) was determined. A polynomial neural network was used, and the model was developed and trained from the collection of 33 pile loading tests. Figure 3.7, shows the performance of the model designed by comparing it with other conventional methods of calculating the lateral capacity of piles. This type of algorithm shows a superior predictive efficiency.

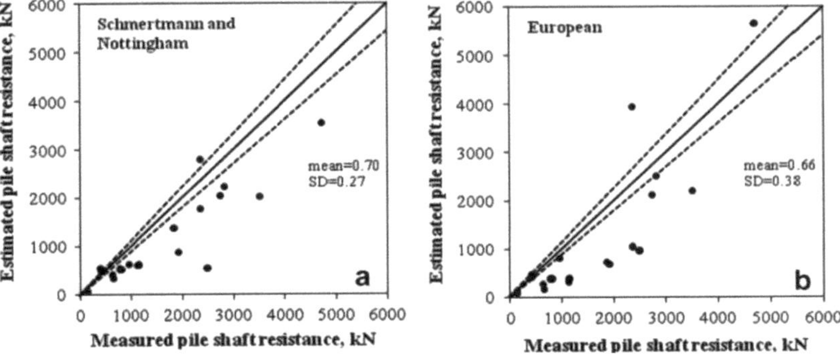

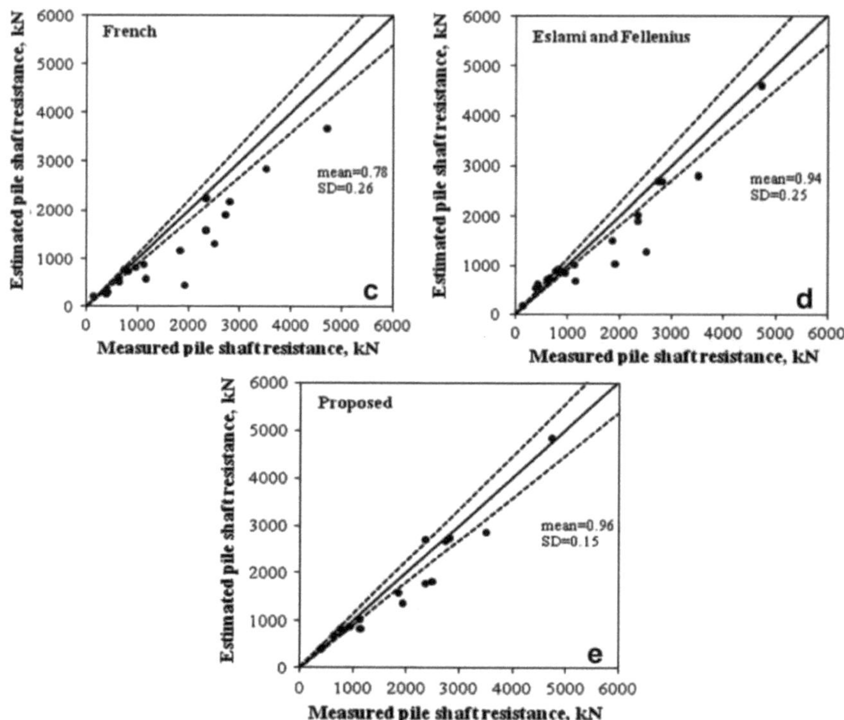

Figure 3.7 Estimated resistances due to lateral friction as a function of those predicted by the various methods [59].

3.3. Integration of neural networks with other computational paradigms to predict the axial capacity of piles

Hung and Adeli (1991b) [61] integrate a genetic algorithm with the backpropagation algorithm within the neural network. The algorithm consists of two learning stages. The first stage aims to accelerate the learning process using the genetic algorithm. In this stage, synaptic weights are encoded in chromosomes as decision variables.

The objective function of the genetic algorithm is defined as the mean square error in the system. After carrying out several iterations and defining the stopping criterion, the first learning stage is complete. The chromosome will return the minimum value of the objective function, which will be considered as the initial weights of the neural network. During this second phase, the backpropagation algorithm performs the second learning process.

Alkroosh and Nikraz (2012) [12] integrated the genetic algorithm and more specifically gene expression programming (Figure 3.8) into the neural network to predict the axial capacity of driven piles in cohesive soils from CPT test data. The model was developed by collecting a database of 25 pile loading cases. Table 3.3, shows the performance of the model by comparing it with other traditional methods and other RNA models for predicting the bearing capacity of driven piles based on the CPT test. The results indicate that the model has a better ability to predict the capacity of driven piles.

McVay et al. (2014) [62] developed a genetic program to predict both pile tip and fut capacities based on the SPT test. The model inputs are: soil type, N_{SPT}, lateral diameter and pile type.

16

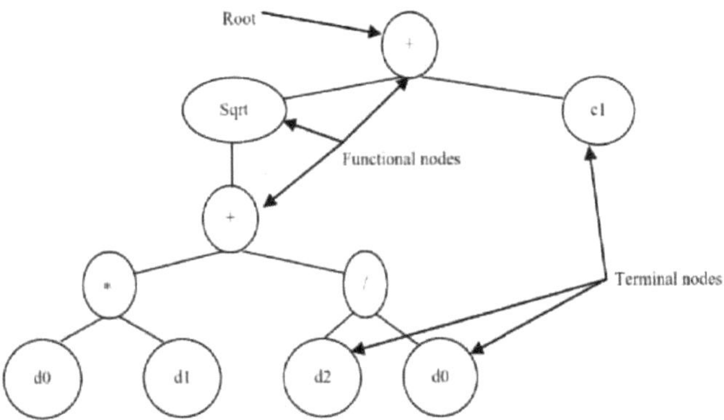

Figure 3.8 Typical example of an expression tree [12].

Table 3.3 Numerical results of the statistical analysis [12].

Methods	Correlation coefficient, r		Average, μ		P*50	
	All data	Validation set	All data	Validation set	All data	Validation set
GEP	0,95	0,94	0,96	1,09	1,01	1,02
De ruiter and Beringen (1979)	0,73	0,73	1,15	1,14	1,05	1,01
Bustamante and Gianeselli (1982)	0,80	0,97	0,70	0,68	0,74	0,68
Eslami and Fellenius (1997)	0,85	0,91	0,98	0,98	1,02	0,95
Shahin (2010)	0,76	0,99	0,83	0,96	0,86	1,05

P^*_{50} = 50% probability, GEP is the model developed

A very limited number of works concern the integration of fuzzy logic with neural networks in the field of deep foundations. *Rajasekaran et al (1996)* [63] describe the integration of fuzzy logic with neural networks in the problem of diagnosing anomalies in prestressed concrete piles. *Ni et al (1996)* [64] used this combination to assess the stability of natural slopes, taking into account geology, topography, meteorology and environmental conditions which were interpreted in linguistic terms.

3.4. The properties of geo-materials

The form of the relationship between soil parameters is very complex, *Goh (1995)* [65] used NRs to model the correlation between relative density and cone strength (the CPT test) for normally consolidated and over-consolidated sands.

Successful NR models using soil parameters as input and the compressibility index as output ([66]-[68]) show that NR-based models can accurately estimate soil swelling.

Park and Kim (2010a) [69] proposed a model for predicting the unconfined stress of reinforced soils (Figure 3.9) (RLS: soil consisting of washed soil, cement, thermo-bubble and fishing waste). This stress is very sensitive to the ratio of the various constituents of the mixture, and as it is difficult to formulate empirically a mathematical relationship between the stress and the constituents of the mixture, the author has developed a RNs model that predicts the non-confining stress of this material for different percentages of the constituents. To do this, laboratory tests were carried out for different percentages of constituents in the mixture

(Figure 3.10a). Fig 3.10b shows a good correlation with the learning and validation of the model developed. The model was able to capture the complex behaviour between stress and the different ratios of materials in the mix.

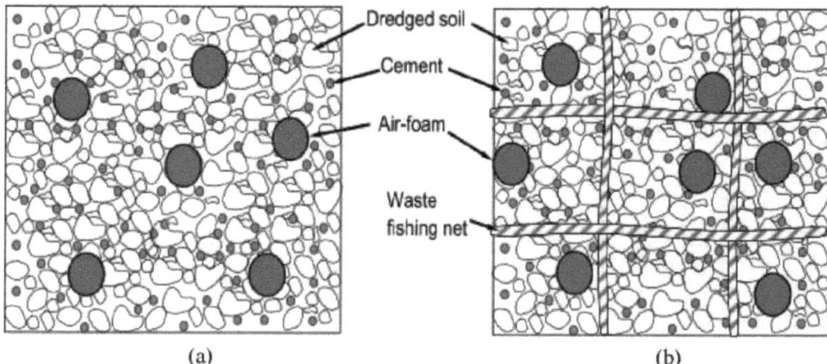

Figure 3.9 Schematic diagram of (a) unreinforced soil and (b) reinforced light soil [69].

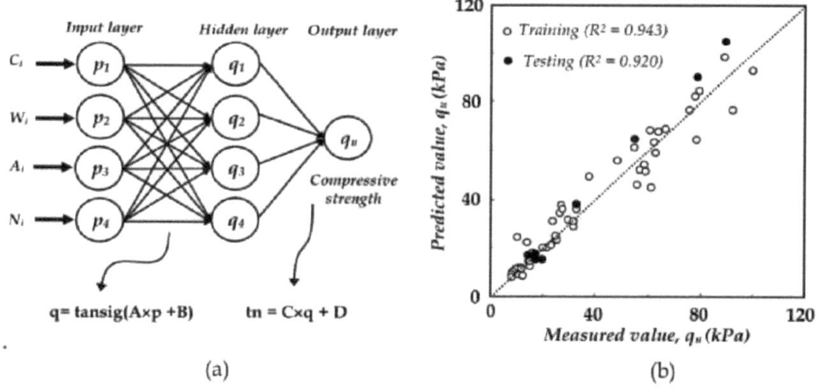

Fig 3.10 Architecture of the neural network model developed [69].

3.5. Slope stability

Slope stability is an important subject and a challenge for geotechnical engineers. Prevention of incidents that may occur in the event of loss of stability or sliding requires an assessment of the process that governs the behaviour of slopes. *Cho 2009*[70] *and Wang et al. 2005*[71] have successfully developed models for assessing slope stability and the appropriate safety factor. This requires geometric, physical, geological and mechanical data on the materials and the existing pore pressure.

3.6. Foundation settlement

The problem of estimating foundation settlement is very complex and uncertain. This has led authors to develop prediction models. *Sivakugun (1998)* [72] developed an RNs model to predict the settlement of shallow foundations in granular soils. The network is trained with 5 inputs: applied pressure, number of average SPT blows, foundation width, foundation shape and foundation depth. Settlement is the model output. The results of the model simulations were compared with those of the calculation methods generally used and produced good results.

18

Shahin and Jaksa (2005) [73] developed a calculation code (RNs model) to estimate the settlement of shallow foundations using 150 cases. Comparison with traditional methods shows better estimation efficiency. The RNs model was used to design settlement tables based on contact stress, footing size and the number of blows of the standard 'N-SPT' penetration test. In designing their model, the authors included the spatial variability of the SPT test at each site.

Pooya et al (2009) [74] developed an RNs model to predict pile settlements, based on the SPT test. For this purpose, 100 cases were collected. The inputs to the model were the axial stiffness of the pile, the cross-sectional area at the base, the perimeter, the length and the anchorage length of the pile. The authors divided the anchorage length into five segments of equal thickness. After correcting for the number of blows, they calculated the arithmetic mean of N in each segment. These five N are introduced as input variables.

3.7. Soil properties and behaviour

The development of correlations between different soil parameters was discussed by *Goh (1995)* [65]. He designed a model that correlates relative density and resistance given by the CPT test. The model is trained with the results of laboratory tests. Model inputs include relative density and average effective soil stress, and the output is penetrometric strength. The model gives a good value for the correlation coefficient, which is of the order of 0.97.

Penemadu and Jean Lou (1997) [75] developed a model for the behaviour of sands and clays. *Penemadu and Zhao (1999)* developed a model to characterise the stress-strain and volume change behaviour of sands and gravels under drained triaxial compression.

Cal (1995) [76] developed a model that generates a quantitative classification of soils based on three factors: plasticity index, liquidity limit and percentage of existing clays.

3.8. Liquefaction

Liquefaction is a phenomenon that occurs in loose, saturated sands during a seismic movement. The soil will lose its shear strength due to the increase in pore pressure. *Goh(1994)* [77] uses RNs to model the relationship between soil parameters and seismic parameters in order to measure the potential for soil liquefaction. The trained model uses 13 seismic data collected in Japan and the USA. The model has 8 inputs. The output takes the value 1 if the site is highly liquefiable and 0 if it is weakly liquefiable.

3.9. Support structures

Goh et al (1995) [78] developed a RNs model which provides an initial estimate of maximum retaining wall deflections for braced excavations in clay soils. The data used to develop this model came from a finite element analysis of braced excavations in clay soils. The maximum deflection is the only output of the model. The results show a good simulation of the RNs model.

3.10. Tunnels and underground openings

Shi et al (1998) [79] presented an RNs model for predicting tunnel settlement. The model has 10 inputs and 3 outputs representing the three settlements: settlement at the passage face, settlement downstream of the passage and settlement after stabilisation.

4. Improving the performance of neural network models in geotechnical engineering

Improving the performance of RNA models requires a systematic analysis of all the stages of model design, such as: the appropriate determination of model inputs, data splitting and pre-processing, the choice of the most suitable architecture, the choice of internal model parameters that best control the optimisation method, the stopping criterion and model validation (Figure 4.1). For example, and in relation to the choice of method for data partitioning (Figure 4.1, step 2), Shahin et *al.* (2004)[80] provide guidelines by recommending the use of three statistically compatible but independent datasets, one for training, the other two for testing and validation. The authors introduce the following three approaches to data division including trial and error, self-organising maps, and fuzzy clustering. A detailed explanation of these approaches can be found in Shahin et *al* (2008)[81].

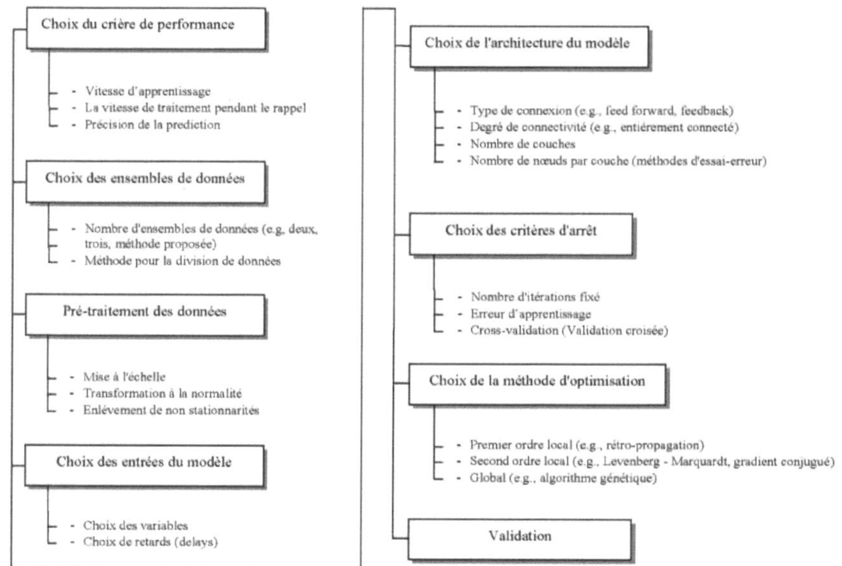

Figure 4.1 Main stages in the development of an RN model [82].

Application of* the *Bayesian technique to train the RNs model

Goh et al (2005) [83] found that by incorporating the Bayesian technique into the back-propagation algorithm, the predictive skill increased, thus lowering the predictive uncertainty of the model. Recent work has shown that Bayesian neural networks provide a very satisfactory model in the field of prediction (Shahin et *al.* 2005) [84], Benali et al 2017[10]).

Other limitations and appropriate solutions

Extrapolation by the RNAs model is as known not possible ([85] ; [86]). This major limitation in the operation of RNs models represents a challenge for researchers. In this context, Sudheer et *al* (2003) [87] have proposed a methodology based on the *'Wilson-Hilferty'* transformation which will enable the model to predict extreme values while respecting the critical value of the river or lake discharge. Recent advances in the application of RNs in geotechnical engineering have involved combining them with other artificial intelligence techniques. One example is the incorporation of fuzzy logic into RN models. This type of model uses fuzzy logic to store or retain knowledge acquired from input variables and their corresponding output variables. The rules of fuzzy linguistics can easily interpret them. Applications of these models in geotechnical engineering include the work of Ni et *al* (1996) [88] and Padmini et *al* (2008) [89].

The combination of the RNA technique with genetic programmes has been discussed and applied by several researchers (e.g., [90]; [12]-[14]). These models have made it possible to obtain a simplified formulation of certain geotechnical problems.

4.1. The neural approach and the implementation of evolutionary and swarm intelligence methods

Very recently, and given the difficulties associated with mathematical learning algorithms (backpropagation), researchers have thought about developing other alternatives, such as modern population-based heuristic methods for finding the optimal solution. Researchers have considered developing other alternatives, such as modern population-based heuristic methods for finding the optimal solution. These can be divided into two groups: evolutionary and swarm intelligence algorithms. Evolutionary methods include genetic algorithms and artificial immunity algorithms. Genetic algorithms are widely used. They work on the principle of Darwin's theory of living beings and their evolution. Particle swarm optimization (PSO) [91] and artificial bee colony (ABC) [92] are techniques from the swarm intelligence family of algorithms. PSO is based on the principle of the behaviour of a group of birds, while ABC works on the behaviour of a bee.

Armaghani et al (2015) [93] integrated the PSO algorithm to adjust the RNs free parameters to predict the axial capacity of piles. Momeni et al (2014) [94] developed a hybrid RNs-GA system to predict the axial capacity of piles based on dynamic pile loading test results. The results their simulation was able to give a coefficient of determination R^2 of 0.99 and an error (MSE) of 0.002. Chattergee et al (2016) [95] developed a RNs system combined with the PSO algorithm for the prediction of failure in reinforced concrete constructions. Choubineh et al (2017) [96] developed a hybrid RNs-TLBO model to predict critical fluid flow rates for oil well production, while Taheri et al(2016) [97] simulated ground vibration produced by dynamite used in mines by a hybrid ANN-ABC model.

CHAPTER 2
Behaviour of an insulated pile under axial loading

1. Introduction

Deep foundations offer major advantages for civil engineering construction. They allow service forces to be channelled to the layer with sufficient mechanical characteristics to take up the forces. The forces will be transferred by two mechanisms: one is lateral friction mobilised along the shaft, the other is peak resistance. Despite major advances in the field, the design of deep foundations remains a difficult problem, linked to complex behavioural mechanisms that are still relatively poorly understood. For this reason, the design of piles is still often linked to the use of calculation methods based on experience (static loading tests, penetrometric and pressuremeter tests) or on empirical methods. In the following section, we present some of the literature that was required to guide our literature review. Particular attention will be paid to installation effects. Some methods for calculating the bearing capacity of piles are then mentioned, in particular the calculation of lateral friction and peak resistance.

2. Installation of an insulated pile
2.1. Effects of installing a pile

The method of installing a pile can have a very significant effect on the pulverulent soil in the vicinity of the pile (deformation, displacement, densification) and on the mechanical response of the pile (bearing capacity). The deformation measurement technique is based on sophisticated imaging equipment [98]. The analysis showed that the shape of the deformation path reveals high vertical compression under the tip followed by horizontal compression as the soil migrates towards the pile shaft [98] (Fig 2.1). The influence of the installation method on the mechanical response of the pile has been studied by several authors [98]. When model piles were tested in a calibration chamber in sand, they observed that cast piles developed less resistance. Flemming et al (1992) [99] estimated the friction mobilised by a bored pile at 70% of that mobilised by a driven pile. De Gennaro (1999) [100] compared the responses of dark model piles to different confining stresses with those of a cast model pile. He found that the cast pile mobilised less peak resistance and lateral friction and a reduction in head limit load values of approximately 50% in the cast pile compared with the dark pile [98].

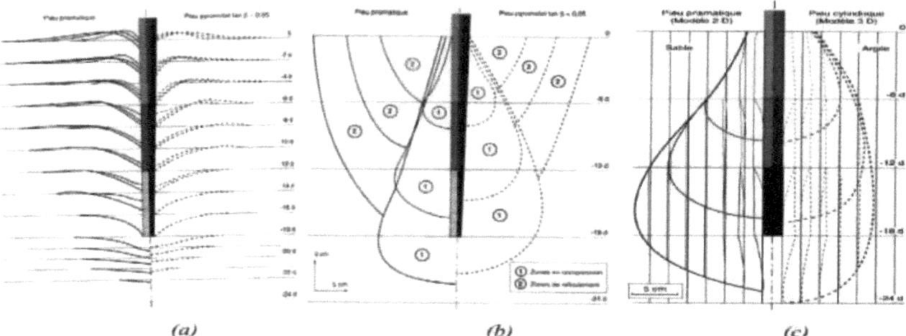

(a)   *(b)*   *(c)*

Figure 2.1 Soil deformation zones when driving pile models in sand (a) vertical displacements (b) compacted and displaced soil zone (c) horizontal displacements [98].

3. Behaviour of axially loaded piles

In the case of piles working in compression, the bearing capacity is given by :

$$Q_t = Q_p + Q_s \tag{3.1}$$

With : Q_t is the total bearing capacity; while Q_p is the capacity due to peak resistance, Q_s is the capacity due to lateral friction (Fig 3.1), where the weight of the pile is neglected. Friction is considered positive when the shear stress is directed upwards (Fig. 3.1).

22

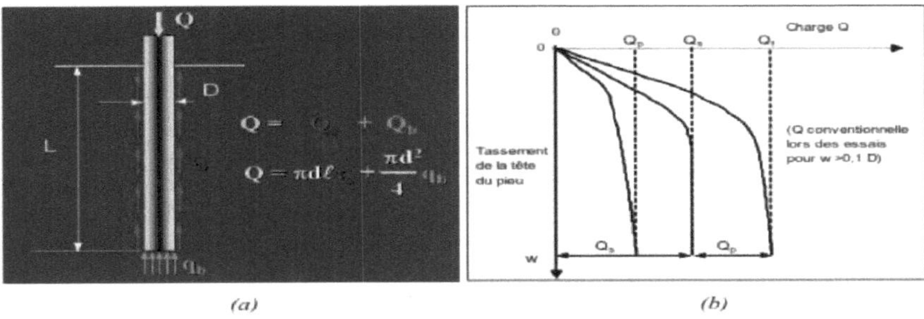

Figure 1.14 (a) Bearing capacity of a pile in compression **(b)** Load mobilisation of an axially loaded pile [98].

3.1. Massif density

The table below gives a few orders of magnitude of the influence of the initial density of the sand through the evolution of the earth pressure coefficient K as well as the installation method and the roughness of the pile (Table 3.1).

Table 3.1 Influence of mass density on the earth pressure coefficient [98].

	Type of pile	K (Loose sand)	K(Dense sand)
FOND (1972)	Steel pile	0,5	1
	Rough concrete pile	1	2
	Smooth concrete pile	0,5	1
	Tapered wooden stake	1,5	4
Puech et *al* (1979)	Type of pile	K (Dr = 20%)	K (Dr = 70%)
	Moulded pile	1,5	3,8
Eissautier (1986)	Type of pile	K (Dr < 30%)	K (Dr > 70%)
	Beaten pile	2 à 3	2 à 3
	Bored pile	0,75 à 1,5	1 à2

2.2. Roughness of the pile interface

The roughness of the pile has a significant effect on frictional behaviour. Figure 3.2 clearly shows that a smooth surface results in low resistance to lateral friction and low stiffness compared to a rough surface.

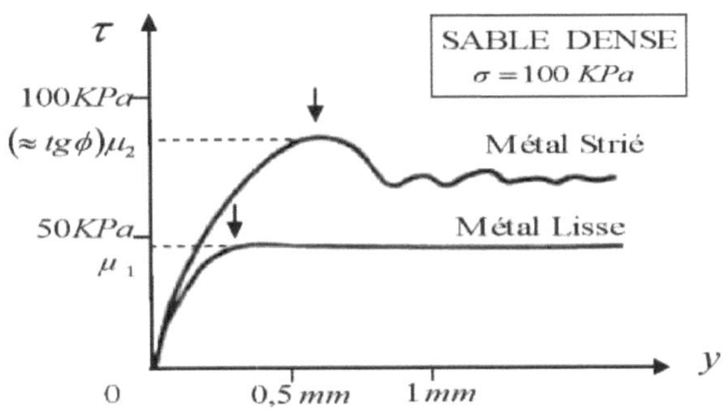

Figure 3.2 Effect of inclusion roughness: Comparison of friction-displacement mobilisation curves [98].

4. Pile design based on SPT test results

4.1. General

In general, methods for calculating the axial capacity of piles using in situ tests fall into two categories.

(1) Indirect methods, or

(2) Direct methods

Indirect methods require the evaluation of characteristic soil parameters, such as the angle of internal friction and undrained shear stress from the results of in situ tests. This requires the consideration of boundary values for this complicated problem (Campanella et *al.* 1989). On the other hand, with the direct methods, the results of the measurements from the in situ tests can be used for the analysis and design of the foundations without the evaluation of the characteristic parameters of the soil. The application of direct methods for the analysis and design of foundations is, however, based on empirical or semi-empirical correlations. Figure 3.3, shows some examples of valid methods for direct and indirect approaches in different applications.

Indirect methods for pile design include: [101]; [44] and [37] for powdery soils, and those of [102]; [37] for cohesive soils. Most indirect methods define correlation factors between the stress state and the strength at the tip and along the shaft of the pile based on soil parameters.

Direct methods are essentially based on the Standard Penetration Test (SPT) and the Static Penetrometer Test (CPT). The SPT test is widely used in the design of deep foundations. The main aim of the following is to estimate the two bearing capacity resistance terms (that at the base of the pile and the other along the shaft) of an axially loaded pile with reference to direct methods based on the SPT, followed by an assessment of the predictive quality of these methods.

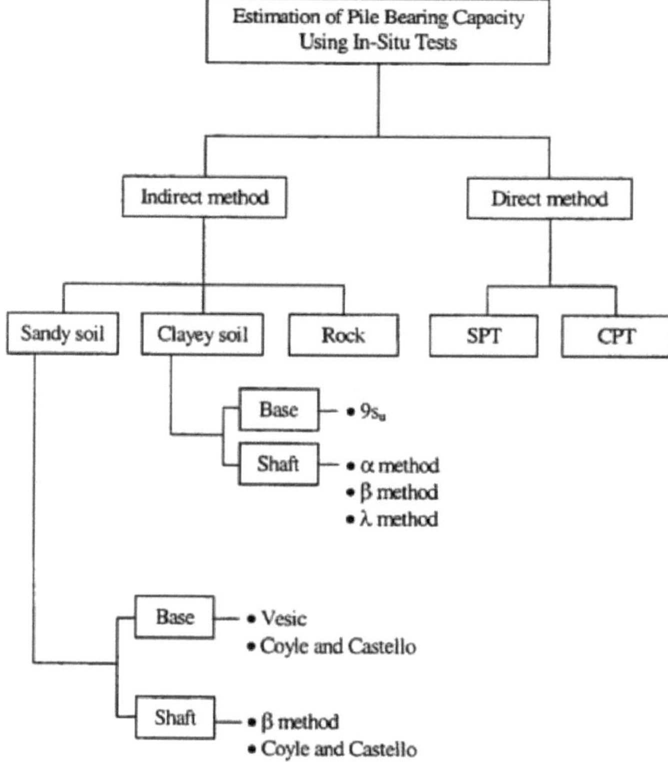

Figure 3.3 Examples of methods for estimating pile bearing capacity [103].

4.2. Estimated bearing capacity of the pile based on the results of the SPT test

In this section, existing methods for calculating the bearing capacity of piles based on SPT will be reviewed. In most SPT-based methods, the axial capacity of piles is defined in terms of the number of blows, together with correlation parameters. These relationships typically have the following form [104]

$$q_b = K_b . N_b \tag{4.1}$$

$$q_s = \sum K_{si} . N_{si} \tag{4.2}$$

API (2008) adopts these methods to predict the ultimate capacity of pipe piles.

Where: qb *is* the pile tip resistance, *Kb* is the conversion factor from number of blows to tip resistance, qs is the resistance along the shaft (due to lateral friction), *Ksi* is the conversion factor from number of blows to lateral resistance (lateral friction), *Nsi* is the number of blows representative of the value of N along the shaft in layer i.

4.2.1. The Meyerhof method

Meyerhof (1976, 1983) [43] proposed the following expressions for peak strength from the SPT test for sands and gravels:

$$q_b = 0.4 . N_{1,60} . D / B . P_a \leq 4 . N_{1,60} . P_a \tag{4.3}$$

Where: N_{60} is the average number of blows corrected for the effect of earth pressure and the effect of the test procedure and ranges from 8 times the diameter above the point to 4 times the diameter below, B is the diameter of the pile, D is the anchorage of the pile, Pa is the reference pressure = 100 KPa.

The upper limits of peak strength given by equation [1.3] are often applied in the case where D/B>10 for sand and gravel. For pile diameters ranged between 0.5<B/ *Br* < 2, WHERE: Br: reference length= 1m, q_b is reduced using the factor *rb* as follows:

$$rb = \frac{B + 0,5.Br}{2} . B \leq 1 \tag{4.4}$$

Where: n = 1, 2 or 3 for loose, medium or dense sand respectively. Meyerhof (1976, 1983) also proposed the expression of resistance along fut for large and small displacement piles

For low displacement piles in non-cohesive soils

$$qs = N_{60} . \frac{Pa}{100} \tag{4.5}$$

For large displacement piles in non-cohesive soils

$$qs = N_{60} . \frac{Pa}{50} \tag{4.6}$$

Where: N_{60} is the number of moves corrected for the effect of the test procedure only

It should be noted that Meyerhof's method does not take into account the effect of these two types of pile installation (wide and low displacement) on peak resistance. Paik and Salgado (2003) conducted a study on two pipe piles in sand, of the same dimensions, one with an open base (small displacement), the other with a closed base (large displacement). They found that the peak resistance of the two piles was different. Meyerhof does not use any criteria to interpret the ultimate load, but he does give a specification of the failure zone. Consequently, a study by Shariatmadari *et al* (2008) [105] showed that Meyerhof's method gives reasonable prediction values.

4.2.2. The Canadian Foundation Engineering Manual

The Canadian Foundation Engineering Manual [106] in its 4ième edition and after revision suggested:

$$Qult = m . N . At + n . \bar{N} . As \tag{4.7}$$

Where: *Qult* is the ultimate axial capacity of the pile in granular soil in KN, *m* is an empirical factor (equal to 400 KPa for driven piles and 120 KPa for bored piles), *N* is SPT is the number of blows at the base of the pile, *n* is an empirical coefficient (equal to 2 KPa for driven piles and 1 KPa for bored piles), *At* is the transverse surface area at the base of the pile in m^2, A_s: the lateral surface area of the pile in m2, N the average number of blows at the base of the pile, corrected by the expression CN. N where,

$$CN = 0,77.\log_{10}\left(\frac{2000}{\sigma'v}\right), \quad \sigma'v \geq 25KPa, \quad \sigma'v \text{ en KPa} \tag{4.8}$$

No specification introduced in the formulation which shows the two installation modes (large and small displacement) and does not use any criteria to interpret the ultimate load and the failure zone.

4.2.3. The Aoki and Velloso method

Aoki and Velloso (1975) [107] proposed the following formula for soil types and piles,

$$qsi = Nsi.Pa.\alpha.\frac{K}{F2} \tag{4.9}$$

$$qb = Nb.Pa.K/F1 \tag{4.10}$$

Where: K is an empirical factor depending on the soil type, Fl and $F2$ are empirical factors depending on the pile type,^{K} is the lateral friction resistance factor depending on the soil type, Nb is the average of three values of NSPT in the vicinity of the pile tip, Ns is the average value of NSPT along the pile shaft in layer i, excluding those used to calculate Nb. The values of K,^{K}, and Fl, $F2$ are shown in Table 4.1(a and b).

Table 4.1(a) Values of K and K, for different soil types [103].

Type of soil	K	K (%)
Sand	10,0	1,4
Silty sand	8,0	2,0
Clayey silty sand	7,0	2,4
Clayey sand	6,0	3,0
Silty clay sand	5,0	2,8
Silt	4,0	3,0
Sandy loam	5,5	2,2
Clayey sandy loam	4,5	2,8
Clay loam	2,3	3,4
Sandy clay loam	2,5	3,0
Clay	2,0	6,0
Sandy clay	3,5	2,4
Sandy silty clay	3,0	2,8
Silty clay	2,2	4,0
Silty sandy clay	3,3	3,0

Table 4.1(b) Values of Fl, F2 for different types of pile [103].

Types of piles	Fi	$F2$
Franki piles	2,50	5,0
Metal piles	1,75	3,5
Prestressed concrete piles	1,75	3,5
Bored piles	3,0 - 3,50	6,0 - 7,0

The use of the SPT test in clay is not recommended. The authors have adopted the Vander Veen failure criterion to interpret the ultimate load. The failure zone is not specified. These factors may lead to inadequate values.

4.2.4. The Briaud and Tuker method

Briaud and Tucker (1984) [108] developed a method for determining the resistance at the pile tip and along the shaft. A hyperbolic variation is adopted for both resistances as a function of pile settlement, as shown below:

$$qb = qres + \frac{S}{\dfrac{1}{kp} + \dfrac{S}{qs\max - qres}} \tag{4.11}$$

$$qs = \frac{S}{\dfrac{1}{kr} + \dfrac{S}{qs\max - qsres}} \tag{4.12}$$

Where,

$$kp = 18684.Npt^{0,0065\frac{Pa}{Br}} \tag{4.13}$$

$$q\max = 19{,}75.Npt^{0,36} \tag{4.14}$$

$$qres = 5{,}57.L.\Omega.Pa \tag{4.15}$$

$$Kr = 200.Nside^{0,27} \tag{4.16}$$

Where: Br is the reference length and is equal to 1m, L is the length of the pile, NSPT is the average uncorrected number of blows in an area which lies between 4 times the diameter below the base of the pile and 4 times the diameter above, Nside is the average corrected number of blows along the shaft. These expressions are valid for driven piles.

This method adopts the failure criterion, which states that the failure load corresponds to a settlement of 1/10 of the pile diameter. A failure zone is defined by the authors. The study by Shariatmadari et al (2008) [105] also showed that the Briaud and Tuker method gives acceptable values.

4.2.5. Robert's method

Robert (1997) [109] proposed the following formulae:

For driven piles For bored piles

$$qb = 1{,}9.\,Pa.\,N_{1,60} \qquad\qquad qb = 1{,}15.\,Pa.\,N_{1,60} \tag{4.17}$$

$$qs = \frac{1{,}9}{1000}.\,Pa.\,\overline{N} \qquad\qquad qs = \frac{1{,}9}{1000}.\,Pa.\,\overline{N} \tag{4.18}$$

Where: N is the average number of blows along the pile, N_{1160} is the number of blows at the base of the pile, noting that N is corrected for effects due to earth pressures.

Robert's method does not specify any criteria or failure zones. It is a general formula proposed by the author for any pile anchored in any type of soil.

4.2.6. The PHRI Standard method (1980)

PHRI Standard [110] suggests the following empirical relationships:

$$qb = 4.\,Pa.\,Nspt \tag{4.19}$$

$$qs = \frac{2}{1000}.\,Pa.\,\overline{N} \tag{4.20}$$

Where: N is the average number of blows along the pile, NSPT is the number of blows in the vicinity of the base of the pile. Note that N is corrected for water table effects. This method applies to driven piles.

4.2.7. The Shioi and Fukui method

Shioi and Fukui (1982)[111] suggested the following expressions

For driven piles.   For bored piles

$$qb = 3.\,Pa.\,Nb \qquad\qquad qb = 1.\,Pa.\,Nb \tag{4.21}$$

$$qs = \frac{2}{1000}\,Pa.\,\overline{N} \qquad\qquad qs = \frac{1}{1000}\,Pa.\,\overline{N} \tag{4.22}$$

Where: N is the average number of blows along the pile, Nb is the number of blows at the base of the pile.

Shioi and Fukui consider neither a failure zone nor a failure criterion. The peak strength of the pile is strongly affected by this zone. Consequently, the errors induced will lead to unreasonable capacity values.

4.2.8. The Reese and O'Neill method

The method focuses on bored piles. Reese and O'Neill (1989)[112] suggested the following expression:

$$qb = 0,6.\,Pa.\,Nb \tag{4.23}$$

$$qs = \frac{3,3}{1000}\,Pa.\,\overline{N} \tag{4.24}$$

Where: N is the average number of blows along the pile, Nb is the average number of blows over a distance of twice the diameter below the tip of the pile.

4.2.9. The Bazaraa and kurkur method

Bazaraa and Kurkur (1989) [113] proposed the following expression for bored piles:

$$qb = 1,35.\,Pa.\,Nb \tag{4.25}$$

$$qs = \frac{0,67}{1000}\,Pa.\,\overline{N} \tag{4.26}$$

Where: N is the average number of blows along the pile, Nb is the average number of blows over a distance of 3.75 times the diameter above the tip of the pile and 1 times the diameter below.

In conclusion, the shortcomings deduced from these methods are summarised as follows: Most of the methods cited above ignore the pressure of the water generated during the test, and the problem is persistent when dealing with soils with low or no permeability, such as clays. The SPT test is generally recommended for sandy or non-cohesive soils.

All the methods either specify a limited breaking zone or do not take it into account. This factor strongly affects the calculated capacities, and this last point must be analysed carefully.

In all the methods based on the SPT test the calculated mean is arithmetic, whereas the curve of variations in N may contain peaks and troughs which may lead to false interpretations. A study carried out by Eslami and Fellenius (1997) [114], showed that the adoption of the arithmetic mean is far from representative of the real case of soil conditions along the pile shaft or in the vicinity of the tip.

5. Evaluation of prediction methods based on the SPT test

Bouafia and Derbala (2002) [115] carried out a study of the statistical performance of the most widely used SPT-based prediction methods, using a database of 46 cases of pile loading tests in predominantly sandy soils. Table 5.1 summarises the results of this analysis. ^ is defined as the average of the ratios of predicted to measured capacity. SD and COV represent respectively the mean standard deviation and the mean coefficient of variation (which is the ratio of μ_{moyen}/SD). The level of under-prediction is defined as the percentage of under-predicted cases. Nine methods were evaluated. One of the best-known semi-empirical methods, based on the SPT test, is that of [37].

From Table 5.1, the coefficient of variation is almost the same for all the methods, so it can be concluded that these methods are characterised by the same level of dispersion. Figure 5.1 shows that the methods of Aoki and Vellosos [107] and Decourt [116] overestimate the bearing capacity in all the cases studied. In accordance with the previous section, this can be explained by the relatively high values of the two bearing or correlation factors, with an average ^ value equal to 1.91.

The Meyerhof and CFEM (1994) method underestimates the load-bearing capacity in most cases, with

28

an average value $\wedge$ equal to 0.71. These pessimistic values may justify the wide distribution and use of this method. The lateral resistance or correlation factor proposed by Meyerhof for bored piles is the lowest. Thus the number of blows in this study is subject to a correction due to depth. As shown in Figure 5.2, the methods of Robert, Bazaraa and Kurkur, Lopes and Laprovitera, Hansen and Burland respectively, seem to give the best predictions with an average $\wedge$ value ranging between 0.93 and 0.98. In fact, two-thirds of the cases analysed are under-predicted by these methods, beyond which they are judged to be pessimistic.

Assuming that the histograms of the number of frequencies of $\wedge$ can be fitted by a Gaussian distribution and according to the ranking criterion defined as the number of frequencies of |a fitting between 0.8 and 1: the methods respectively of Robert, Lopes and Laprovitera, Bazaraa and Kurkur, Burland and Hansen are characterised by frequencies of: 30.8%, 28.9%, 24.3% and 21.5% (Bouafia and Derbala 2002[115]). However, all the methods tested express a certain level of dispersion in the predicted values, with a mean value of 30% (Figure 5.2).

In this context, it was considered appropriate to carry out an evaluation of the methods for calculating the bearing capacity of piles based on the SPT test. The analysis will be carried out on a data set of 40 cases selected from different sources and sites, including publications as well as experimental programmes. The details of this study are illustrated in the next section.

Table 5.1 Evaluation of SPT-based capacity calculation methods [115].

Methods	Average □	Min	Max	SD.	VOC % (%)	Level of under-prediction(%)	Level of over-prediction(%)
Bazaraa and Kurkur (1986)	0.97	0.35	1.92	0.31	31.8	58.0	42.0
Decourt (1982)	3.13	1.29	6.55	0.99	31.6	0.0	100.0
Lopes and Laprovitera (1988)	0.95	0.37	2.40	0.32	33.2	68.2	31.7
Meyerhof (1976) CFEM (1985)	0.70	0.31	2.03	0.25	36.2	94.4	5.6
Shioi et al (1982)	0.82	0.36	1.86	0.27	31.6	83.2	16.8
Aoki and Velloso (1975)	1.91	0.86	4.27	0.60	31.3	7.5	92.5
Reese et al (1989)	1.11	0.43	3.08	0.40	36.0	50.5	49.5
Robert (1997)	0.98	0.44	2.56	0.34	35.1	66.3	33.7
Hansen-Burland (1973)	0.93	0.32	1.54	0.30	32.3	67.3	32.7

Note. SD: Mean standard deviation; COV: Coefficient of variation

5.1. Predictive evaluation of load-bearing capacity calculation methods based on the SPT test
5.1.1. Database used

The database of cases extracted from the results of 40 pile loading tests are compiled with information on soil type and SPT test results carried out in the vicinity of the pile location points. These cases are obtained from different sources and, reporting data from several sites in several countries. Table 5.2 summarises the main characteristics of this test. The soil is made up of different soil profiles, ranging from coherent to powdery. Only most of the soil profiles in this set are predominantly sandy. Concrete piles, either driven or drilled, come in a variety of shapes and sections. This database will be the subject of a comparative study between the most commonly used methods for calculating the bearing capacity of piles subjected to compressive loads, based on the SPT test.

Table 5.2 Description of the database used [15].

Sources	Rentals	Number of cases	Geometric characteristics of the piles
1	USA,	15	B = 0.1 to 1.22 m; D =

	Canada, Germany		2.25 to 57.1 m ; D / B = 11.9 to 74.26 Total = 40
2	Malaysia	1	
3	Texas	1	
4	Kuwait	1	
5	Malaysia	1	
6	France	1	
7	Texas	3	
8	Bangkok	10	
9	Malaysia	4	
10	Las Vegas	3	

1:Abu Keifa (1998) 2:Balakrishnan et al (1999); 3:Reese and O'Neill (1971); 4:Ismael (2001);5:Amaludin and Hussein(1998); 6:Bustamante and Gianeselli (1980); 7:Briaud et al (2000); 8:Thasnanipan et al (1998); 9: Abdul Aziz and Lee(2005); 10: Mackiewicz and Lehman (2004).

5.1.1. Performance criteria used

The methods used to carry out this assessment are as follows: Meyerhof (1976)/CFEM(1994); PHRI Standard(1980); Aoki and Velloso(1975); Shariatmadari (2008), Shioi et al (1982); Bazaraa and Kurkur (1986), Briaud and Tucker (1988), Reese et al (1989), Robert (1997) (Table 1.8). The optimum performance of the various methods of predicting the ultimate capacity of the pile is indicated by the following two criteria:

- The first criterion concerns the determination of certain statistical parameters such as the arithmetic mean (μ_{LL}) and the standard deviation (o).
- The second criterion is evaluated by plotting the Log-Normal distribution of the $Q_{p/Qm}$ ratio for the methods concerned by this evaluation. Based on the results of the analysis of the Log-Normal distribution of the selected methods, the probability (P) that the predictions of the ultimate capacity are within an accuracy or error level of ±25% is obtained by calculating the area of the Log-Normal distribution within this interval.

5.1.2. Results and discussion

The calculation results according to the chosen performance criteria are summarised in Table 5.3 and Table 5.4.

According to the first criterion (Table 5.3), the methods of Shariatmadari, Meyerhof, Briaud & Tucker and Robert give the best results for driven piles. The statistical parameters, which are the coefficient of determination R^2 and the mean $\wedge$ of the ratio of the predicted load to the measured load, have the following values: 0.93 and 1.15 for Shariatmadari; 0.86 and 0.97 for Meyerhof; 0.89 and 0.92 for Briaud and Tucker; 0.90 and 0.78 for Robert. The Shioi and Fukui method appears to give the weakest parameters, with a coefficient of determination R^2 of 0.26 and an average of 0.87. For bored piles, the methods of Bazaraa & Kurkur, Shioi and Fukui respectively rank first with a coefficient of determination R2 and an average of: 0.73 and 1.07 for Bazaraa & Kurkur; 0.73 and 0.80 for Shioi and Fukui.

The second criterion (Table 5.4 and Figure 5.3) is evaluated by plotting the Log-Normal distribution of the $Q_{p/Qm}$ ratio for the methods of: Shariatmadari (2008), Meyerhof (1976), Reese et *al* (1989), Aoki and Velloso (1975), Bazaraa and Kurkur (1986) and de Robert (1997). Based on the results of the analysis of the Log-Normal distribution of the selected methods, the probability (P) that the predictions of the ultimate bearing capacity fall or occur within an accuracy or error level of ±25% is obtained by calculating the area of the Log-Normal distribution within the interval, $0.75Q_m < Q_p < 1.25Q_m$. Based on this criterion, the highest probability implies adequate accuracy of the prediction method. The results obtained show that the induced error of the Shariatmadari, Meyerhof and Robert calculation methods is within an acceptable range. These methods predict the bearing capacity of a pile subjected to compression with reasonable accuracy. The Shariatmadari, Meyerhof and Robert methods are the most suitable for calculating the bearing capacity of driven piles, while the Bazaraa & Kurkur method appears to be the most suitable for simulating bored piles.

Table 5.3 Performance of load-bearing capacity calculation methods based on the SPT test

Methods	Types of piles	Performance parameters		
		R2	H	S
	Beaten	0.86	0.97	0.30
Meyerhof (1976)/ CFEM 1994	Foré	0.68	0.71	0.15
PHRI Standard (1980)	Beaten	0.78	1.14	0.35
Shioi and Fukui (1982)	Beaten	0.26	0.87	0.27
	Foré	0.73	0.80	0.31
Bazarra and Kurkur (1986)	Foré	0.73	1.07	0.36
Reese et *al* (1989)	Foré	0.63	1.30	0.34
Robert (1997)	Beaten	0.90	0.78	0.27
	Foré	0.74	0.80	0.26
Briaud and Tucker (1988)	Beaten	0.89	0.92	0.35
Shariatmadari et *al* (2008)	Beaten	0.93	1.15	0.41
Aoki and Velloso (1975)	Beaten	0.81	0.77	0.40

Note: n is the mean c is the mean standard deviation

Table 5.4 Probability of estimation for calculation methods with an error of ±25%.

Methods chosen	Types of piles	Probability of estimation with a percentage error of ±25% in(%)
Aoki and Velloso (1975)	Beaten	25
Meyerhof (1976)/CFEM (1985)	Beaten	36
Bazaraa and Kurkur (1986)	Foré	27
Reese et al (1989)	Foré	19
Shariatmadari et al(2008)	Beaten	59
Robert (1997)	Beaten	40

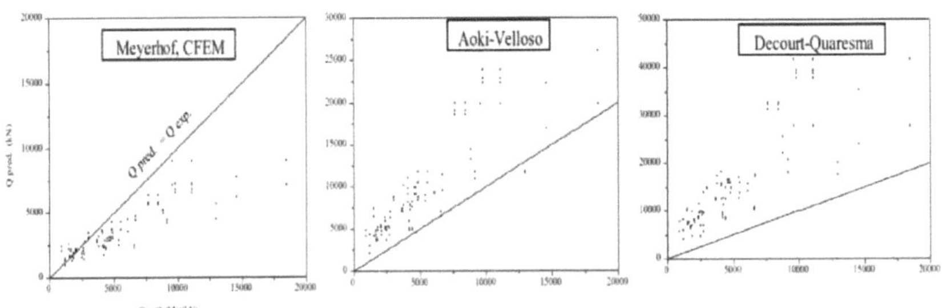

31

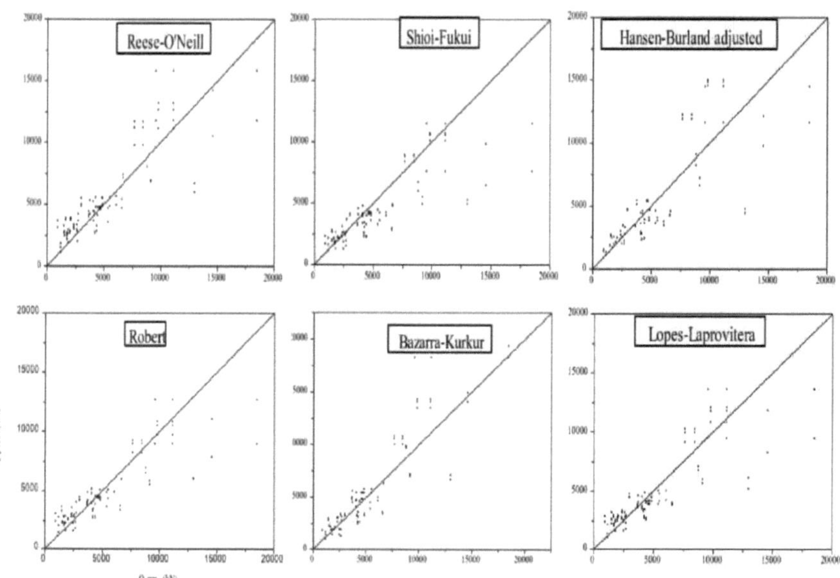

Figure 5.1 Comparison of experimental values with those given by SPT-based methods [115].

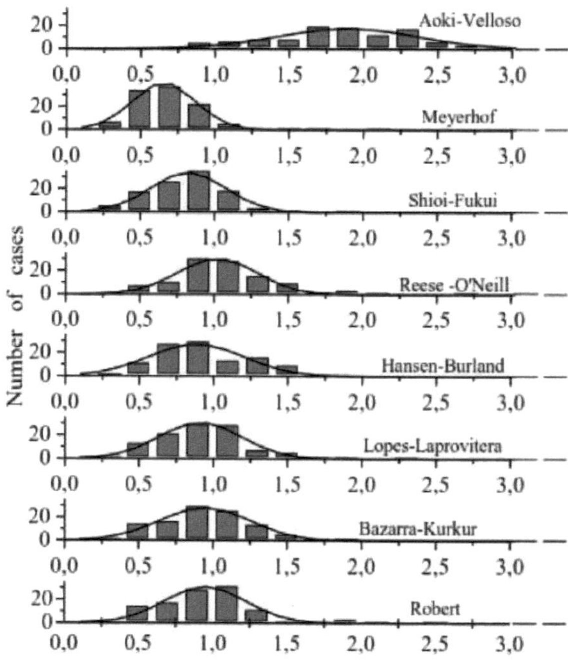

Figure 5.2 Histograms of the u frequency of the different calculation methods [115].

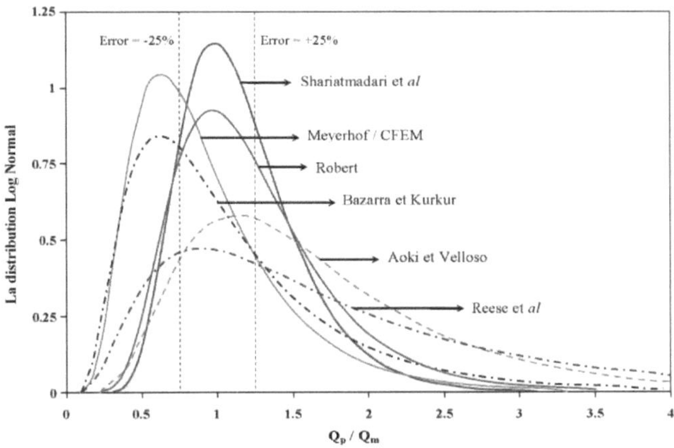

Figure 5.3 Log Normal distribution of the selected calculation methods

References

[1] . Benali, A.: Analyse semi-empirique de la portance des pieux isolés (in French). Postgraduate dissertation (Magister degree), University of Blida, Algeria (2002).

[2] . Iskander, M. 2011. Behavior of pipe piles in set: Plugging and Pore-Water Pressure Generation during Installation and Loading, Geomechanics and Geoengineering, Springer Berlin Heidelberg Edition. 250p.

[3] . Bandini, P.; Salgado, R. 1998. Methods of pile design based on CPT and SPT results, Proceeding of the 1st International Conference on site characterization, Balkema, Rotterdam: 967-976.

[4] . Shariatmadari, N.; Eslami, A.; Karimpour-Fard, M. 2006. Bearing capacity of driven piles in sets from SPT-applied to 60 case histories, Iranian Journal of Science and Technology, Transaction B, Engineering 32(B2): 125-140.

[5] . Abu Kiefa, M. A. 1998. General Regression neural networks for driven piles on cohesionless soil, Journal of Geotechnical and Geoenvironmental Engineering 124(12): 1177-1185

[6] . Shahin, M. A. 2014. Load settlement modeling of axially loaded drilled shafts using CPT based recurrent neural network, Soils and Foundations 54(3): 515-522.

[7] . Kordjazi, A.; Pooya Nejad, F.; Jaksa, M. B. 2014. Prediction of ultimate axial load carrying of piles using vector machine based on CPT data, Computer and Geotechnics 55: 91-102.

[8] . Azizkandi, S. A.; Kashkooli, A.; Baziar, M. H. 2014. Prediction of uplift pile displacement based on cone penetration tests (CPT), Gotechnical and Geological Engineering 32(4): 10431052.

[9] . Mohammad, H. B.; Azizketi, S. A.; Kashkooli, A. 2015. Prediction of pile settlement based on cone penetration test results: An ANN approach, KSCE Journal of Civil Engineering 19(1): 98-106.

[10] . Benali, A.; Boukhatem, B.;Hussien, M. N.; Nechnech, A.; Karray, M., 2017 Prediction of axial capacity of piles driven in non-cohesive soils based on neural approximation, Journal of Civil Engineering and Management, 23(3), 393-408

[11] . Boukhatem, B.; Kenai, S.; Tagnit-Hamou, A.; Ghrici, M. 2011. Application of New Information Technology on Concrete: An Overview. Journal of Civil Engineering and Management, International Research and Achievements, Edition Taylor & Francis, 17(2):248- 258.

[12] . Alkroosh, I.; Nikraz, H. 2012. Predicting axial capacity of driven piles in cohesive soils using intelligent computing, Journal of Engineering Application of Artificial Intelligence 25: 618-627.

[13] . Ismail, A.; Jeng, D. S.; Zhang, L. L. 2013. An optimized product- unit neural network with a novel PSOBP hybrid training algorithm: Applications to load-deformation analysis of axially loaded piles, Engineering Application of Artificial Intelligence 26(10): 2305-2314.

[14]   . Ahangar-Asr, A.; Javadi, A. A.; Johari, A.; Chen, Y. 2014. Lateral load capacity modeling of piles in cohesive soils in undrained conditions: An intelligent evolutionary approach, Applied Soft Computing 24: 822-828.

[15]   . Benali, A.: Application of artificial neural networks to predict axial bearing capacity of piles. Phd thesis, University of Science and Technology, Algiers, Algeria (2016).

[16]   . Kamp, Y. Hasler, M. 1990. Recursive Neural Networks for Associative Memory. Presses Polytechnique et Universitaires Romandes, Lausanne, 231p.

[17]   . McCulloch, W. S. and Pitts, W. A. 1943. A Logical Calculus in the Ideas Imminent in Nervous. Activity, Bulletin of Mathematical Biology, Vol.5, pp.115-133.

[18]   . [18] Hebb, D.O. 1949. The Organization of Behavior. Wiley, New -York. USA, 37p.

[19]   . [19] Rosenblatt, F. 1958. The Perceptron: A probabilistic Model for Information Storage and Organization in the Brain. Psychological Review, Vol.65, pp.386-408.

[20]   . Widrow, B. and Hoff, M. E. 1960. Adaptive Switching Circuits. IRE WESCON Convention Record, Part 4, pp.96-104.

[21]   . Minsky, M. Papert, S. 1969. Perceptrons, Cambridge MIT. Press.

[22]   . Hopfield J.J. 1982. Neural Network and Physical Systems with Emergent Collective Computational Abilities, Proceedings of the National Academy of Science, pp.2554-2558.

[23]   . Grossberg, S. 1973. Contour Enhancement, Short-trem Memory, and Consistencies. In. Reverberating Neural Network, Statically Application and Mathematical, Vol.52, pp.217- 257.

[24]   . Yeddou, Y. M. 1998. Etude de Synthèse sur les Réseaux de Neurones et leurs Applications. Magister thesis, Ecole Polytechnique d'Alger, 150p.

[25]   . Korn, H. 1997. The Unexpected in Neurophysiology. Pour la Science, Vol.235, pp.10-13.

[26]   . Renders, J .M. and Nairn, P. 1995. Genetic Algorithms and Neural Networks. Ed. Hermes, Paris, France, 334p.

[27]   . Orr   M.   J.   L.   1996.   Introduction   to   Radial   Basis   Function   Networks. http://www.anc.ed.ac.uk/~mjo/napers/intro.ps.gz.

[28]   . Davalo, E. and Naim, P. 1993. Les Réseaux de Neurones. Ed. Eyrolles, 220 p.

[29]   . Jodouin, J.F. 1994. Les Réseaux de Neurones : Principe et Définition", Ed. Hermes, 75017, Paris, 320 p.

[30]   . Alliot, J.M. and Schiex, T. 1993. Les Réseaux de Neurones. In Intelligence Artificielle et Informatiques Théorique, Ed. Cepadues Toulouse, France, p.421-440.

[31]   . Rumelhart, D. E. Hinton, G. E. and Williams R. J. J. 1986. Learning Internal Representations by Error Propagation. In. Parallel Distributed Processing, Vol.1, Cambridge, MA, MIT Press, pp.318-362.

[32]   . Dreyfus, G.; Martinez, J. M.; Samuelides, M.; Gordon, M. B.; Badran, F.; Thiria, S.; Herault, L. 1994. Réseaux de neurones - Méthodologie et application, Edition Eyrolles.

[33]   . Chung, Y. & Kusiak, A. (1994). Grouping parts with a neural network, Journal of Manufacturing Systems, Vol.13, No.4, pp. 262-75.

[34]   . Kusiak, A. & Lee, H. (1996). Neural computing based design of components for cellular manufacturing, International Journal of Production Research, Vol.34, No.7, pp. 1777-1790.

[35]   . Goh, A. T. C. 1994a. Nonlinear modeling in geotechnical engineering using neural networks, Australian Civil Engineering Transactions CE36 (4): 293-297.

[36]   . Semple, R. M.; Rigden, W. J. 1986. Shaft capacity of driven pipe piles in clay, Ground Engineering 19(1): 11-17

[37]   . Burland, J. B. 1973. Shaft friction of piles in clay, Ground Engineering 6(3): 1-15.

[38]   . Goh, A. T. C. 1996b. Pile driving records reanalyzed using neural networks, J. Geotech. Engrg, ASCE 122(6): 492-495.

[39]   . Chan, W. T.; Chow, y. K.; Liu, L. F. 1995. Neural network: An alternative to pile driving formulas. J Computers and Geotechnics: 17: 135-156

[40]   . Broms, B. B.; Lim, P. C. 1988. A simple pile driving formula based on stress-wave measurements. Proc. 3rd Int. Conf. on the Application of Stress-Wave Theory to Piles, B. H. Fellenius, ed, Vancouver,

591-600.

[41]   . Rausche, F.; Moses, F.; Goble, G. G. 1972. Soil resistance predictions from pile dynamics. J Soil Mech. and Found. Div, ASCE 98: 917-937

[42]   . Lee and Lee (1996).

[43]   . Meyerhof, G. G.1976. Bearing capacity and settlement of pile foundations, ASCE Journal of Geotechnical Engineering 102(3):1-19.

[44]   . Coyle, H. M.; Castello R. R. 1981. New design correlations for piles in set, Journal of the Geotechnical Engineering Division ASCE 127(GT7): 965-986.

[45]   . API (American Petroleum Institute. 1984.

[46]   . Randolph, M. F. 1985. Capacity of piles driven into dense set, Report. Soils TR 171, Eng. Dept, Cambridge University, Cambridge, U. K.

[47]   . Narendra, B. S.; Sivapullaiah, P. V.; Suresh, S. N. Omkar. 2006. Prediction of unconfined compressive strength of soft grounds using computational intelligence techniques: a comparative study, Computers and Geotechnics 33(3): 196-208.

[48]   . Park and Cho (2010).

[49]   . Teh, C.I.; Wong, K.S.; Goh, A.T.C.; Jaritngam, S. 1997. Prediction of pile capacity using neural networks, J Comput Civ Eng, ASCE 11(2): 129-138.

[50]   . Rausche, F.; Moses, F.; Goble, G. G. 1972. Soil resistance predictions from pile dynamics. J Soil Mech. and Found. ASCE 98: 917-937.

[51]   . Kuo et al (2009)

[52]   . Shahin, M.A. 2010. Intelligent computing for modeling axial capacity of pile foundations, Canadian Geotech Journal 47(2): 230-243.

[53]   . MacKay (1992a, 1992b).

[54]   . Adeli, H.; Hung, S. L. 1995. Machine Learning: Neural Networks, Genetic Algorithms, and Fuzzy Systems, Wiley, New York.

[55]   . Carpenter, W. C.; Barthelemy, J. F. 1994. Common misconceptions about neural networks as approximators, Journal of Computing in Civil Engineering, ASCE 8 (3): 345-58.

[56]   . Adeli, H.; Hung, S. L. 1994. An adaptive conjugate gradient learning algorithm for effective training of multilayer neural networks, Applied Mathematics and Computation 62 (1): 81102

[57]   . Yeh, I. C. 1998. Structural engineering applications with augmented neuron networks, Computer-Aided Civil and Infrastructure Engineering 13(2): 83-90.

[58]   . Adeli, H.; Hung, S. L. 1994. An adaptive conjugate gradient learning algorithm for effective training of multilayer neural networks, Applied Mathematics and Computation 62 (1): 81102.

[59]   . Ardalan, H.; Eslami, A.; Nariman-Zadeh, N. 2009. Piles shaft capacity from CPT and CPTu data by polynomial neural networks and genetic algorithms, J Comput Geotech 36: 616-625.

[60]   . Ivakhnenko, A. G. 1971. Polynomial theory of complex systems, IEEE Trans Syst Man Cybern, SMC 1: 364-78.

[61]   . Hung, S. L.; Adeli, H. 1991b. A hybrid learning algorithm for distributed memory multi computers, Heuristics, Journal of Knowledge Engineering 4 (4): 58-68

[62]   . McVay, M. C.; Klammler, H.; Tran, K. 2014. Pile/Shaft design using artificial neural networks (i. e., Genetic Programming) with spatial variability considerations, Final Report: 10/15/12-5/15/14, Contract No. BDK 75-977-68, University of Florida, Dept. of Civil and Costal Engineering, 114P, USA.

[63]   . Rajasekaran, S.; Febin, M. F.; Ramasamy, J. V. 1996. Artificial fuzzy neural networks in civil engineering, Computers and Structures 61 (2): 291-302.

[64]   . Ni, S. H.; Lu, P. C.; Juang, C. H. 1996. A fuzzy neural network approach to evaluation of slope failure potential, Computer Aided Civil and Infrastructure Engineering 10(1): 59-66.

[65]   . Goh (1995).

[66]   . Ozer, M.; Isik, N.S.; Orhan, M. 2008. Statistical and neural network assessment of the compression index of clay-bearing soils, Bull Eng Geol Environ 67: 537-545.

[67]   . Park, H. I.; Lee, S.R. 2010. Evaluation of the compression index of soils using an artificial neural

network, Computers and Geotechnics.

[68] . Najjar, Y.M.; Basheer, I. A. 1996. Utilizing computational neural networks for evaluating the permeability of compacted clay liners, Geotechnical and Geological Engineering 14: 193-221

[69] . Park, H. I.; Kim, Y.T. 2010. Prediction of strength of reinforced lightweight soil using an artificial neural network, Engineering Computation.

[70] . Cho, S. M. 2010. Foundation design of the Inchon Bridge, Geotech. Eng. J of the SEAGS and AGSSEA 41(4).

[71] . Wang, H. B.; Xu, W.Y.; Xu, R. C. 2005. Slope stability evaluation using Back propagation neural networks, Engineering Geology 80: 302- 315

[72] . Sivakugun, .1998.

[73] . Shahin, M. A.; Jaksa, M. B. 2005. Neural network prediction of pullout capacity of marquee ground anchors, Computers and Geotechnics 32(3): 153-163.

[74] . Pooya et al .2009.

[75] . Penemadu, and Lou, J. 1997.

[76] . Cal (1995).

[77] . Goh, A. T. C. 1994a. Nonlinear modeling in geotechnical engineering using neural networks, Australian Civil Engineering Transactions CE36 (4): 293-297

[78] . Goh, A. T. C. 1995b. Empirical design in geotechnics using neural networks, Geotechnique, 45(4): 709-7 14

[79] . Shin, S. W.; Yun, C. B; Futura, H.; Popovics, J. S. 2008. Non-destructive evaluation of crack depth in concrete using PCA-compressed wave transmission function and neural networks, Experimental Mechanics 48: 225-231.

[80] . Shahin, M. A.; Maier, H. R.; Jaksa, M. B. 2004. Data division for developing neural networks applied to geotechnical engineering, Journal of Computing in Civil Engineering, 18(2): 105-114

[81] . Shahin, M. A.; Jaksa, M. B.; Maier, H. R. 2008. State of the art of artificial neural networks in geotechnical engineering, Electronic Journal of Geotechnical Engineering, 8: 1-26

[82] . Benali, A. 2016. Application of artificial neural networks to predict axial bearing capacity of piles. Phd thesis, University of Science and Technology, Algiers, Algeria.

[83] . Goh, A. T. C.; Kulhawy F. H.; Chua, C. G. 2005. Bayesian neural network analysis of undrained side resistance of drilled shafts, Journal of Geotechnical & Geoenvironniental Engineering 131(1): 84-93.

[84] . Shahin, M. A.; Jaksa, M. B.; Maier, H. R. 2005. Stochastic simulation of settlement of shallow foundations based on a deterministic neural network model, In Proceedings of the International Congress on Modeling and Simulation (MODSIM '05): 73-78, Melbourne, Australia.

[85] . Flood, I.; Kartam, N. 1994. Neural networks in civil engineering: I) principles and understating, Journal of Computing in Civil Engineering 8(2): 131-148

[86] . Tokar, A. S.; Johnson, P. A. 1999. Rainfall-run of modeling using artificial neural networks, Journal of Hydrologic Engineering 4(3): 232-239

[87] . Sudheer, K. P.; Nayak, P. C.; Ramasastri, K. S. 2003. Improving peak flow estimates in artificial neural network river flow models, Hydrological Processes 17(3): 677-686

[88] . Ni, S. H.; Lu, P. C.; Juang, C. H. 1996. A fuzzy neural network approach to evaluation of slope failure potential, Computer Aided Civil and Infrastructure Engineering 10(1): 59-66

[89] . Padmini et al. .2008.

[90] . Alkroosh, I.; Shahin, M. A.; Nikraz, H. R. 2008. Modeling axial capacity of bored piles using genetic programming technique, In Proceedings of the 3rd International Geotechnical Conference 113-120, Shinghai, Thailand.

[91] . Kennedy, J. and Eberhart, R.C. 1995. Particle swarm optimization, Proceedingsof IEEE International Conference on Neural Networks, IEEE Press, Piscataway, 1942-1948.

[92] . Karaboga, D. and Bastrurk,B. .2008. On the performance of artificial bee colony algorithm. Applied Soft Computing, 8(1), 687-697.

[93] . Armaghani D. J and Faizi, K.2015.Developing a hybrid PSO-ANN model for estimating the ultimate

bearing capacity of rock socketed piles. Neural Comput and Applic.DOI 10.1007/s00521-015-2072-z

[94] . Momeni, E. Nazir, R. Armaghani, D. J and Maizir, H. 2014. Prediction of pile bearing capacity using a hybrid genetic algorithm-based ANN. Measurement, 57, 122-131.

[95] . Chatterjee1, S. Sarbartha, S. Hore, S. Dey, N. A. S.and Balas, V. E. 2016. Particle swarm optimization trained neural network for structural failure prediction of multistoried RC buildings. Neural Comput and Applic. DOI 10.1007/s00521-016-2190-2

[96] . Choubineh, A. Ghorbani, H. Wood, D. A. Moosavi, S. R. Khalafi, E. Sadatshojaei E. 2017. Improved predictions of wellhead choke liquid critical-flow rates: Modelling based on hybrid neural network training learning based optimization. Fuel, 207, 547-560.

[97] . Taheri1, KH. Hasanipanah, M. Golzar, S. B. Abd Majid, M.Z. 2016. A hybrid artificial bee colony algorithm- artificial neural network for forecasting the blast~ produced ground vibration. Engineering with Computers. DOI 10.1007/s00366-016-0497-3

[98] . Said, I. 2006. Comportement des interfaces et modélisation des pieux sous charges axiales, Ph.D thesis, Ecole Nationale des Ponts et Chaussées, Paris, France.

[99] . Flemming, W. G. K; Weltman, A. J.; Randolph, M. F.; Elson, W. K. 1992. Piling Engineering, 2nd Edition, Jon Wiley and Sons, New York, N.Y.

[100] . De Gennaro, V.; Frank, R. 2002 b. Insight into the simulation of calibration chamber tests, Proc. Eur. Conf on Num. Methods in Geotechnical Engineering (NUMGE 2002), Mestat Ed., Presses de l'ENPC/LCPC, Paris, 169-177.

[101] . Vesic, A. S. 1977. Design of pile foundations, Transportation Research Council, Washington, D.C.

[102] . Bowles, J. E. 1988. Foundation analysis and design, McGraw-Hill BookCompany, Singapore.

[103] . Salgado, R.; Lee, J. 1999. Pile design based on cone penetration test results, Joint Transportation Research Program, Indiana Department of Transportation, Purdue University. 242 p.

[104] . Bandini, P.; Salgado, R. 1998. Methods of pile design based on CPT and SPT results, Proceeding of the 1st International Conference on site characterization, Balkema, Rotterdam: 967-976.

[105] . Shariatmadari N, Eslami A, Karimpour-Fard M. 2008. A new method for estimation the bearing capacity of piles based on SPT results. 31st DFI Annual Conference on Deep Foundations, Washington, DC.

[106] . Canadian foundation engineering manual. 1992, 4th Edition, Canadian geotechnical society, Technical Committee on Foundation, 512 pp.

[107] . Aoki, N.; Velloso, D. 1975. An approximate method to estimate the bearing capacity of piles, Proceedings of the 5th Pan-American Conference on soil Mechanics and Foundation Engineering, Buenos-Aires.

[108] . Briaud, J. L.; Tucker, L. M. 1988. Measured and predicted axial capacity of 98 piles. J. Geotechnical. Engineering, ASCE 114(9): 984-1001

[109] . Robert, Y. A. 1997. Few comments on pile design, Can Geotech J, 34: 560-567

[110] . PHRI. 1980. Technical standards for port and harbor facilities in Japan. Bureau of Ports and Harbors, Ministry of Transport, Japan.

[111] . Shioi, Y.; Fukui, J. 1982. Application of N-value to design of foundation in Japan. In: Proceeding of the 2ᵈ ESOPT, Amsterdam 1: 159-164.

[112] . O'Neill, M.; Ata, A.; Vipulanetan, C.; Yin, S. 2002. Axial performance of ACIP piles in Texas Coastal soils, Deep Foundations, ASCE. Geotechnical Special Publication, II(116): 1290-1314

[113] . Bazaraa, A. R.; Kurkur, M. M. 1986. N-values used to predict settlements of piles in Egypt. Proceedings of In Situ'86, New York, 462-474.

[114] . Eslami, A.; Fellenius, B. H. 1997. Pile capacity by direct CPT and CPTu methods applied to 102 case histories, Canadian Geotechnical Journal 34: 886-904.

[115] . Bouafia, A.; Derbala, A. 2002. Assessment of SPT-based methods of pile bearing capacity- analysis of a database, In Proceedings of the International Workshop on Foundation Design Codes and Soil Investigation in View of International Harmonization and Performance-based Design, 369-374.

I want morebooks!

Buy your books fast and straightforward online - at one of world's fastest growing online book stores! Environmentally sound due to Print-on-Demand technologies.

Buy your books online at
www.morebooks.shop

Kaufen Sie Ihre Bücher schnell und unkompliziert online – auf einer der am schnellsten wachsenden Buchhandelsplattformen weltweit! Dank Print-On-Demand umwelt- und ressourcenschonend produziert.

Bücher schneller online kaufen
www.morebooks.shop

Printed by Books on Demand GmbH, Norderstedt / Germany